Studies in Systems, Decision and Control

Volume 82

About this Series

The series "Studies in Systems, Decision and Control" (SSDC) covers both new developments and advances, as well as the state of the art, in the various areas of broadly perceived systems, decision making and control- quickly, up to date and with a high quality. The intent is to cover the theory, applications, and perspectives on the state of the art and future developments relevant to systems, decision making, control, complex processes and related areas, as embedded in the fields of engineering, computer science, physics, economics, social and life sciences, as well as the paradigms and methodologies behind them. The series contains monographs, textbooks, lecture notes and edited volumes in systems, decision making and control spanning the areas of Cyber-Physical Systems, Autonomous Systems, Sensor Networks, Control Systems, Energy Systems, Automotive Systems, Biological Systems, Vehicular Networking and Connected Vehicles, Aerospace Systems, Automation, Manufacturing, Smart Grids, Nonlinear Systems, Power Systems, Robotics, Social Systems, Economic Systems and other. Of particular value to both the contributors and the readership are the short publication timeframe and the world-wide distribution and exposure which enable both a wide and rapid dissemination of research output.

More information about this series at http://www.springer.com/series/13304

Émilie Masson · Marion Berbineau

Broadband Wireless Communications for Railway Applications

For Onboard Internet Access and Other Applications

Springer

Émilie Masson
Institut de Recherche Technologique Railenium
Famars
France

Marion Berbineau
French Institute of Science and Technology for Transport, Development and Networks
Villeneuve d'Ascq Cedex
France

ISSN 2198-4182 ISSN 2198-4190 (electronic)
Studies in Systems, Decision and Control
ISBN 978-3-319-47201-0 ISBN 978-3-319-47202-7 (eBook)
DOI 10.1007/978-3-319-47202-7

Library of Congress Control Number: 2016952900

Printed on acid-free paper

This Springer imprint is published by Springer Nature
The registered company is Springer International Publishing AG
The registered company address is: Gewerbestrasse 11, 6330 Cham, Switzerland

Preface

This book was written in the line of the roadmap of Railenium which is one of the 8 Institutes for Innovation, Research and Technology (IRT) created by a French governmental decree to boost economic competitiveness by filling up the gap between academia research and industry. Railenium is the French IRT dedicated to the railway systems, infrastructures and equipments. The missions of Railenium, on behalf and with the commitment of its members, are to achieve research and development projects, training and investments and exploitation of test facilities.

Railenium is structured as a Foundation for scientific cooperation (non-profit organization) and its activities cover the systems, infrastructures and equipment's for conventional, high speed and urban railway systems. Its aim is to bring together the railway expertise from research centers and companies (manufacturers, railway operators and infrastructure managers, engineering companies) to find innovative solutions both to enhance the competitiveness of the industry and the performance of European rail systems and networks. It has been founded in 2012 as a public/private partnership by 28 stakeholders.

Railenium is Associate Member on Innovation Programme 2 (Advanced Traffic Management and Control Systems) of the Joint Undertaking Shift2Rail through the SmartRaCon consortium, composed with the German Research Center DLR, the Spanish Technology Center CEIT and the British Company NSL.

Émilie Masson, researcher at Railenium and Marion Berbineau, senior researcher at IFSTTAR, a funding member of Railenium, have written this survey on all the railway applications requiring broadband wireless communications. The initial objective of the works was to explore the techniques and existing solutions to provide Internet access on board trains. Authors then thought it might be relevant to broaden the subject to all applications requiring broadband communications in the railway context. Reviewers cited below confirm the benefit to regroup such overviews on the subject.

The authors would like to thank warmly Pierre Cotelle, Networks and Telecom Solution Director at Alstom Transport Information Solution and Thomas Chatelet, ERTMS Project Officer at European Railway Agency for having reviewed this book. By their expertise in the railway domain, they brought real added value to the work.

Famars, France Émilie Masson
Villeneuve d'Ascq Cedex, France Marion Berbineau

Contents

Acronyms

ABA	Axle Box Acceleration
ACM	Adaptive Code Modulation
AGV	Automotrice à Grande Vitesse
AGW	Access Gateway
AI	Artificial Intelligence
AP	Access Point
ATC	Automatic Train Control
ATCS	Advanced Train Control System
ATO	Automatic Train Operation
ATP	Automatic Train Protection
ATS	Automatic Train Supervision
AVLS	Automatic Vehicle Location System
BNSC	British National Space Centre
BS	Base Station
BTS	Base Transceiver Station
C3S	Command, Control and Communication System
CAPEX	CAPital EXpenditure
CBTC	Communication Based Train Control
CCTV	Closed-Circuit TeleVision
CDMA	Code Division Multiple Access
CE	Cognitive Engine
CENELEC	European Committee for Electrotechnical Standardization
CEPT	European Conference of Postal and Telecommunications Administrations
COST	Cooperation in Science and Technology
CR	Cognitive Radio
CSI	Channel State Information
DAMA	Demand Assigned Multiple Access
DVB-RCS	Digital Video Broadcast Reverse Channel Satellite
DVB-S	Digital Video Broadcast Standard

DVB-S2	Digital Video Broadcast Standard 2
EIRENE	European Integrated Railway Radio Enhanced NEtwork
EMC	Electromagnetic Compatibility
EPLRS	Enhanced Position and Location Reporting System
ERA	European Railway Agency
ERRAC	European Rail Research Advisory Council
ERRI	European Rail Research Institute
ERTMS	European Rail Traffic Management System
ESA	European Space Agency
ETCS	European Train Control System
ETSI	European Telecommunications Standards Institute
EU	European Union
FAMOUS	FAst MOving USers
FBMC	Filter Band-based MultiCarrier
FCC	Federal Communications Commission
FC	Fog Computing
FFT	Fast Fourier Transform
FMECA	Failure Mode, Effects and Criticality Analysis
FTP	File Transfer Protocol
GEO	Geostationary Earth Orbit
GMSK	Gaussian Minimum Shift Keying
GPRS	General Packet Radio Service
GPS	Global Positioning System
GSM	Global System for Mobile communications
GSM-R	Global System for Mobile communications for Railway
HAP	High Altitude Platform
HSDPA	High Speed Downlink Packet Access
HSPA	High Speed Packet Access
HST	High Speed Train
HSUPA	High Speed Uplink Packet Access
ICOM	Intelligent COMmunication
ICT	Information and Communication Technologies
IEC	International Electrotechnical Committee
IGZ	Industrie Gruppe Zugbus
IoT	Internet of Things
IP	Internet Protocol
ISDN	Integrated Services Digital Network
ISP	Internet Service Provider
ITCS	Incremental Train Control System
IT	Information Technology
ITS	Intelligent Transportation System
ITU	International Telecommunication Union
KPI	Key Performance Indicator
LAN	Local Area Network
LCX	Leaky Coaxial Cable

LDPC	Low Density Parity Check
LED	Light-Emitting Diode
LEO	Low Earth Orbit
Li-Fi	Light Fidelity
LOS	Line-Of-Sight
LTE	Long Term Evolution
M2M	Machine-to-Machine
MAC	Medium Access Control
MCG	Mobile Communication Gateway
MEC	Mobile Edge Computing
MEO	Medium Earth Orbit
MF-TDMA	Multi Frequency Time Division Multiple Access
MIMO	Multiple Input Multiple Output
MNO	Mobile Network Operator
MORANE	Mobile Radio for Railways Networks in Europe
MSS	Mobile Satellite System
MUD	Multi-User Detection
NFV	Network Function Virtualization
NLOS	Non Line-Of-Sight
NRZ	Non-Return-to-Zero
OFDM	Orthogonal Frequency Division Multiplexing
OFDMA	Orthogonal Frequency Division Multiplexing Access
OOK	On-Off Keying
OPEX	OPerational EXpenditure
P2P	Peer-to-Peer
PA	Public Announcement
PAS	Public Address System
PFD	Probability of Failure on Demand
PIS	Passenger Information System
PMR	Professional Mobile Radio
PTC	Positive Train Control
PTS	Positive Train Separation
PU	Primary User
QoS	Quality of Service
QPSK	Quadrature Phase Shift Keying
RAMS	Reliability, Availability, Maintainability, and Safety
RDERMS	Railway Distributed Energy Resource Management System
RIMMS	Railway Integrated Measuring and Monitoring System
RMPA	Reliable Mobility Pattern Aware
RoF	Radio-over-Fiver
SCPC	Single Channel Per Carrier
SDN	Software-Defined Networking
SDR	Software-Defined Radio
SGW	Service Gateway
SIL	Safety Integrity Level

SMS	Short Message Service
SNR	Signal-to-Noise Ratio
SRS	Safety-Related System
SWiFT	Seamless Wireless Internet for Fast Trains
TAT	Train Access Terminal
TCMS	Train Control and Monitoring System
TCN	Train Communication Network
TDMA	Time-Division Multiple Access
TETRA	Terrestrial Trunked Radio
TGNRM	Train-to-Ground Network Reference Model
TVSE	Télévision Semi-Embarquée
TVWS	TeleVision White Space
UDN	Ultra-Dense Network
UIC	Union of Railways
VBS	Voice Broadcast Service
VCTS	Virtually Coupled Train Sets
VGCS	Voice Group Call Service
VLC	Visible Light Communication
VNF	Virtualized Network Function
VoD	Video on Demand
VoIP	Voice over IP
VPN	Virtual Private Network
VSAT	Very Small Aperture Terminal
W-CDMA	Wideband Code Division Multiple Access
Wi-Fi	Wireless Fidelity
WiGig	Wireless Gigabit
WIGWAM	WIreless Gigabit With Advanced Multimedia
WLAN	Wireless Local Area Network
WRAN	Wireless Regional Area Network
WSN	Wireless Sensor Network
WSSUS	Wide-Sense Stationary Uncorrelated Scattering

List of Figures

List of Tables

Introduction

Wireless communications have been deeply integrated into people's life and current public telecommunication services increased the needs of mobility services. In this context, the current works in standardization groups, such as 3GPP (including evolution towards 5G systems) are dealing with the future wireless radio communications able to answer the increasing high demands of mobile phone users in terms of availability, throughput and reliability. Indeed, the multiplication of mobile terminals, such as smartphones and tablets, leads to a need of ubiquitous connection, everywhere and all the time. Thus, the future wireless technologies will be designed in order to create an ecosystem for technical and business innovation. The telecommunication infrastructures will provide network solutions for many domains, including transportation.

These trends are observed also in the railway domain which relies more and more on wireless communications for vital and non-vital applications related to train operation and passenger demands. Several applications are concerned with various needs such as safety, reliability, availability or high capacities.

The European Rail Research Advisory Council (ERRAC) targeted for the year 2020 to double passenger traffic by rail. Such a goal should be achieved reducing costs, enhancing environmental sustainability and offering new services to passengers. For instance, broadband Internet access has become, in recent years, an essential and highly expected service, whatever the time of the day and regardless the location (home, office or public places, transportation). One can observe the development of services such as "remote desktop" or Quality of Service (QoS) of broadband Internet access for a variety of applications such as messaging, Video on Demand (VoD), Voice over IP (VoIP), TeleVision (TV), streaming, videoconferencing, etc.

Ensuring broadband links between trains and infrastructure also allows considering, for network managers and railway industries, applications hitherto difficult to ensure. Some stakeholders talk about "networked trains", able to ensure several applications, such as real-time video surveillance from inside the carriages, or track inspection in direct link with the Control Center by data feedback of measurements and diagnosis.

Several wireless communications can be set up in the railway context, such as communication between equipment, between vehicles, between consists and between train and ground. The book focuses on the wireless communication link between train and ground. The first chapter presents the needs of railway operators in terms of wireless communications found in the literature. All these needs can be established from a complete study of user needs and use cases. The needs can be divided in two main categories: the commercial services to make on board traveling more comfortable and pleasant, such as Internet access and associated applications, and the operational needs, such as surveillance, maintenance and diagnosis. The second chapter focuses on the definitions of all available technologies and combination of technologies that can be used to provide Internet access on board trains. The book details also all the other operational applications requiring high capacity. Finally, the last chapter highlights challenges and trends in railway telecommunications. The future and emerging technologies, such as 5G and Cognitive Radio concept, are presented. The current discussions and works in the different authorities, dealing with telecommunications, railway specifics or professional networks are highlighted. The key challenges and scientific barriers are also discussed.

Chapter 1
Railway Operators Needs in Terms of Wireless Communications

This chapter is dedicated to the railway operators needs in terms of wireless communications, through railway communication standards, safety aspects and classifications of the applications. Then, the chapter details the operational needs for communication and signaling systems or monitoring systems. Finally, the chapter highlights the needs for services to passengers, though the Internet access on board especially.

1.1 Introduction to Wireless Communications for Railway Applications

This section introduces the context of wireless communications between train and ground for railway applications, the railway communication standards and the safety aspects. Then, different classifications are performed to distinguish between the applications, depending on intended applications, required throughputs and railway sectors. Finally, the basic architectures for train-to-ground applications are introduced.

1.1.1 Context

Required railway services lead to the development of railway communication systems. The need to deploy a communication network on board trains appeared in the late seventies, at first for presenting diagnostic information to the driver and maintenance staff to obtain the status of the whole control system with synthetic messages [1]. In the past, applications for signaling and data communications in the railway domain were assured by robust wires that carried information with significant current load and voltage level from 24 to 110 V [1]. Since several years, wireless systems replaced the wired ones [2]. With the development of electronics,

É. Masson and M. Berbineau, *Broadband Wireless Communications for Railway Applications*, Studies in Systems, Decision and Control 82,
DOI 10.1007/978-3-319-47202-7_1

information and telecommunication technologies, the needs for transmission in the railway domain have increased in order to increase travel safety, optimize the use of existing infrastructure by increasing the frequency of trains, reduce operating costs and maintenance, thus reducing the impact of transport on the environment. The needs of transmissions related to the operation and maintenance of trains and tracks also add the needs of information and services to staff and customers at any time. With the increase of transmission needs, we talk about Intelligent Transportation Systems (ITS) to reflect the introduction of electronic and computer devices contributing to the automation of functions. Systems are called "informed" and vehicles "connected" [3].

Furthermore, the European Rail Research Advisory Council (ERRAC), an advisory body to the European Union (EU) Commissions representing Member States and all stakeholders in this sector, targeted for the year 2020 to double passenger and freight traffic by rail [4]. The objective, presented in the Multi-Annual Action Plan of the Shift2Rail Joint Undertaking, is to provide a seamless, integrated and safe high speed passenger service, door-to-door freight service and an efficient metropolitan and urban mass transport [5]. Such a goal should be achieved by reducing costs, enhancing environmental sustainability and offering new services to passengers.

The objectives will then be reached by increasing the number of information exchanges systems between stakeholders. Each functional entity of the global transport system or each application requires information exchanges more or less frequent and more or less consuming in terms of radio spectrum resources. As a consequence, more and more wireless communication devices operating at different frequencies will be deployed inside the trains and along European Railway lines in the next years, to meet all these communication needs. The particular case of High Speed Train (HST) and the fast increase of train speed lead to more and more attention on the issue of train operation safety. From an operational point of view, the track infrastructure, the rolling-stock and the signaling system are the three main parts contributing to safety operation. The signaling system represents the key part and can be seen as the nerve center of the system [6]. Concerning the services to passengers, such as the Internet on board, needs are growing drastically due to the current public telecommunication services, that increase the needs of mobility services.

The specificity of railway environments implies several complex factors to establish a wireless communication from ground to wayside [6, 7]. First of all, the radio propagation channel characteristics depend on the type of geographic environment encountered. The case of a railway environment is characterized by specific scenarios, such as tunnels, cuttings, crossing bridges or railway stations. The widespread of tunnels is especially a barrier to seamless communication due to the need for radio coverage inside the tunnels. Moreover, the high speeds of HST context lead to problems that are not encountered in highway context for instance. Until 200 km/h, wireless channel can be assumed to be Wide-Sense Stationary Uncorrelated Scattering (WSSUS). This WSSUS assumption is no more valid because of the rapid time-varying and non-stationary HST context. Furthermore, the LOS component of the signal exists in a majority of railway scenarios. The spatial correlation between different multipath is then quite important [6].

The other complicating factors are summarized below:

- The possible metallic structure of the train behaves as a Faraday cage causing important losses on transmission signals (one can note that the development of trains in composite materials will modify this);
- Railway environment can be defined as a "high vibration" and "high interferences" environment, which can lead to a need of isolation of communication devices;
- Large temperature variations are observed;
- Railway environment suffers of high electrical stresses: cohabitation between high power (traction) and low power systems (electronic), strong magnetic fields as for the MAGLEV trains and trains not designed to provide a stabilized power supply;
- Railway companies constantly add or remove rail cars from trains. It is necessary that the communication networks discover this automatically.

The major prerequisite condition to guarantee a reliable communication is the knowledge of the wireless channel and its propagation characteristics [6]. Indeed, works and models were largely developed for public land mobile communications but some of them are not suitable for the specific case of railway environments. For instance, it was shown that the conventional Hata model prediction may reach an error of 20 dB in a HST context with cuttings and viaducts [8]. This domain is an extensive domain of publications these recent years. Channel modeling can deal with different kinds of communication: 60 GHz communications [9], V2V communications [10], MIMO communications [11] and the channel models can rely on different techniques such as modeling from channel sounding [12].

Finally, research works are currently on going for the case of Train Control and Monitoring System (TCMS) in the European project Roll2Rail [13]. The aim of the Work Package dedicated to TCMS is to develop a new generation of train communication systems relying on wireless technologies, thus reducing on board communication cables and simplifying train coupling procedures.

1.1.2 Railway Communication Standards

1.1.2.1 Historical of TCN Needs and Specification

As presented in Sect. 1.1.1, basic Train Communication Network (TCN) for TCMS were deployed in the late seventies to be able to satisfy the following specific requirements [1]:

- Suitable to be used in the railway environment (referring especially to climatic and physical conditions);
- Able to be quickly reconfigured when the train composition changes (vehicles coupling and uncoupling), which is a typical railway requirement;
- Oriented to different data scope and performance (there are in general different needs for control signal and diagnosis or monitoring systems);

- Able to support high level of Reliability, Availability, Maintainability and Safety (RAMS, detailed in the Sect. 1.1.3 dedicated to safety aspects).

The interoperability for the first wired systems was guaranteed by the International Union of Railways (UIC). A standardized 13-poles cable was used to distribute basic commands (door closure, light on and off, Public Address System (PAS)).

Then, a process of standardization was performed by coordinated activities between standardization groups, industry and exploitation. The standardization of the TCN is performed by a working group of the International Electrotechnical Committee (IEC) within the Technical Committee TC9. Two specifications are initially set up for TCN standardization: IEC 61375-1 (dealing with TCN specification) and IEC 61375-2 (dealing with TCN conformance testing). In parallel, UIC prepared some interoperability leaflets, such as UIC 556 and UIC 558. Finally, some European Railway industries set up a Joint Development Project (JDP) to specify a new on board network, by collecting experience from industry, such as ABB, AEG, SIEMENS, FIREMA and developing communication modules to fulfill railway needs. Experiments on a real train were set up by a special consortium of railway operators, the Industrie Gruppe Zugbus (IGZ), leaded by the European Rail Research Institute (ERRI). The aim was to verify in a real environment the IEC 61375 standard and the UIC 556 leaflet. Thus, IEC 61375-1 and IEC 31675-2 specifications and UIC 556 and UIC 558 were first published in 1999 after JDP/IGZ works.

For further information, the interested reader can find a state of the art of on-board wired networks in the railway domain in [14].

1.1.2.2 Current Evolution of TCN

The TCN evolution then continued through European projects, such as the ROSIN, the TRAINCOM, the MODTRAIN or the INTEGRAIL projects (some of them will be detailed thereafter). The TCN standardization has also evolved by the set up of the Working Group WG43. The objectives of the IEC TC9 WG43 are:

- to maintain and update the IEC 61375-1 and IEC 61375-2 documents;
- to define and implement a work plan to upgrade and complete the TCN framework, adding new vehicle bus candidates, broadband train network, wireless communication system between train and ground, etc.

The different standards, regularly updated by the IEC TC9 WG43, are summarized in the Table 1.1. In addition, the Fig. 1.1 presents a synthetic view of the different standards IEC 61375 governing the overall TCN system.

The WG43 relies its works on a "bottom-up" approach due to previous standardization works performed by WG42. Indeed, WG42 standardized first the TCN by specifying Wireless Train Backbone (WTB) and Multipurpose Vehicle Bus (MVB) from physical layer. WG43 has extended the standardization to Ethernet Train Backbone (ETB) and Ethernet Consist Network (ECN) and worked also on the stan-

Table 1.1 The different IEC 61375 standards governing the TCN system

Standard	Description	Last update	Stability update
IEC 61375-1	General architecture specification	06/2012	2017
IEC 61375-2-1	Wire Train Bus (WTB)	06/2012	2017
IEC 61375-2-2	WTB conformance testing	06/2012	2017
IEC 61375-2-3	TCN communication profile	07/2015	2017
IEC 61375-2-4	TCN application profile	02/2016	2016
IEC 61375-2-5	Ethernet Train Backbone (ETB)	08/2014	2016
IEC 61375-2-6	On board to ground communication	10/2015	2017
IEC 61375-2-7	Wireless Train Backbone (WTB)	04/2014	2016
IEC 61375-3-1	Multipurpose Vehicle Bus (MVB)	06/2012	2017
IEC 61375-3-2	MVB conformance testing	06/2012	2017
IEC 61375-3-3	CANopen Consist Network (CCN)	06/2012	2017
IEC 61375-3-4	Ethernet Consist Network (ECN)	03/2014	2020

dardization of the communication protocols between train and ground, defining the Mobile Communication Gateway (MCG).

One can note that there is a coordination and parallel voting between the works carried out at international level at IEC and those performed at European level at the European Committee for Electrotechnical Standardization (CENELEC).

1.1.2.3 The IEC 62580 Standard for Railway Multimedia Applications

For the definition of the standard, the multimedia applications are defined as all applications not related to safety functionality. It comprises then all the applications to provide additional supporting services for passengers, operators and crew. The WG46 of the IEC TC9 was set up in 2009 in order to work on railway multimedia applications and ensure interoperability between applications and proper interface with the supporting communication network [15]. The architecture, based on works from INTEGRAIL project, aimed to provide a platform for multimedia products, reducing development costs and enabling interoperability of multimedia applications at service level. Contrary to WG43, WG46 conducts its works using a "top-down"

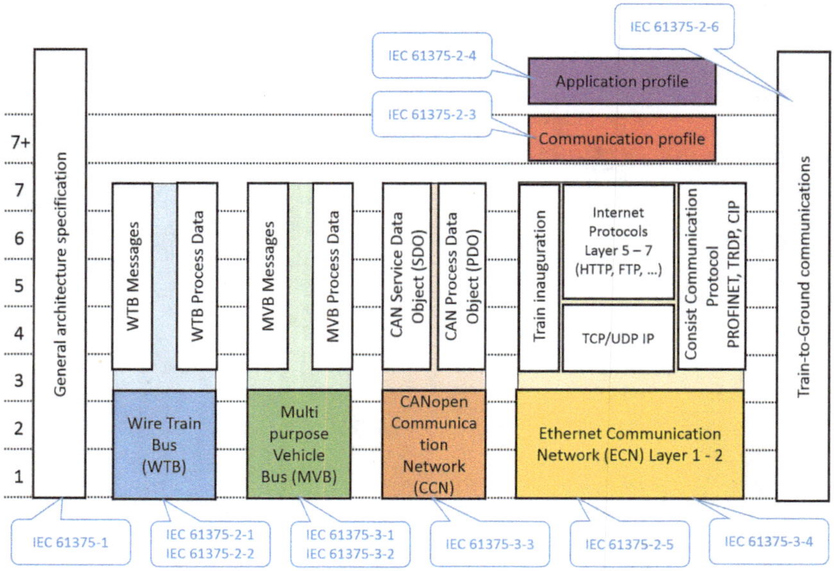

Fig. 1.1 Synoptic of the IEC 61375 standards governing TCN from [1]

approach. Indeed, the objective is to standardize the on board multimedia using a functional and system description to describe the behavior and functional interfaces of the different sub-systems [15]. Coordination between the two WG is obviously essential. At the beginning, an important issue was to clarify the border between the multimedia applications and TCMS.

Works were divided into 5 parts, being supported by subgroups (SG):

1. SG1: General architecture, as a common basis for all multimedia applications, corresponding to the IEC 62580-1 standard;
2. SG2: Security oriented services, like Closed-Circuit TeleVision (CCTV), corresponding to the IEC 62580-2 standard;
3. SG3: Driver and crew oriented services, such as energy management and cab to cab calls, corresponding to the IEC 62580-3 standard;
4. SG4: Passenger oriented services, such as passenger information and seat reservation, corresponding to the IEC 62580-4 standard;
5. SG5: Operator and maintainer oriented services, such as remote monitoring, corresponding to the IEC 62580-5 standard.

The definition of the architecture for all multimedia applications is then defined in the SG1. SG2 works on security oriented services and has to follow the privacy rules and the national laws. Some examples of applications are: rearview mirror (video cameras on the rear of the train to help the driver), outside surveillance (video surveillance of the outside of the train, such as near the doors), inside surveillance (embedded video surveillance of the inside of the train for security applications) or inside listening (audio surveillance of the inside of the train, also for security applications). SG3 deals with the driver and crew oriented services, such as communication between driver and control center or energy management by eco-driving for instance. SG4 interested in passenger oriented services. These ones can be divided in two subjects: the passenger information and the passenger infotainment. The passenger information can include live information on the journey (next stop, current stop, time of arrival) or passenger emergency intercom (communication with crew). The passenger infotainment can consist in Internet access or applications such as VoD. SG5 is responsible for operator and maintainer oriented services, such as remote monitoring and diagnosis, remote maintenance or energy management.

1.1.3 Safety Aspects

Requirements for wireless systems in the railway domain can deeply depend on the criticality of the communication and also on the safety aspects of the system, that are defined in this part.

1.1.3.1 Definitions

Systems whose failure can lead to damage to property, damage to the environment or loss of human life are considered as safety-critical systems [16]. Safety manages design of the system but also the operational environment in which the system is used. Indeed, safety of a system can be dramatically modified by changes in its operational environment. Programmable electronics, often controlled remotely via communication networks, progressively replace the mechanical and mechatronic devices to ensure safety of the devices. Safety-critical systems can no more be based only on the control of errors and failures, they must manage the security of the data used in their operation.

A range of standards relating to the use of electronics components and software in safety-critical systems are defined. These standards aim to design, procure and deploy safety-critical systems that provide some assurance on the safety features of the systems. The IEC 61508 presents a generic approach of all activities dedicated to functional safety of Electrical/Electronic/Programmable Electronic (E/E/PE) devices. The standard proposes a global approach of safety, that can be compared to the ISO 9000 system in the quality domain [17]. The IEC 61508 promotes a design methodology framework that aims to prevent the presence of dangerous failures, or control them when they arise by providing guidance on each phase of the safety life-cycle.

The safety requirements of a Safety-Related System (SRS) must be specified in terms of the functions to be performed by the SRS and the integrity required of each. Each safety-related function is then specified by a Safety Integrity Level (SIL). The SIL is defined by the necessary action to reduce the risk of a function, from an uncontrolled risk to a tolerable risk. The SIL can then quantify the level of security of a system. The level 1 (SIL 1) corresponds to the lower safety integrity level and the level 4 (SIL 4) represents the higher safety integrity level:

- SIL 4: catastrophic impact (higher level);
- SIL 3: impact on the community;
- SIL 2: major protection of installation and production or injury risk on persons;
- SIL 1: minor protection of installation and production (lower level).

1.1.3.2 Application to the Railway Domain

According to these definitions, several systems of the overall railway system can be considered as safety-critical. Safety strategies employed by the railway industry deal with the safe separation of trains and fail-safe protection of paths through junctions and crossings. The separation of trains strategy relies on the assumption of an instantaneously stop of the lead trains and suboptimal breaking performance for the following trains. The fail-safe protection of paths strategy is based on the use of interlockings. Different signaling strategies can be implemented from these assumptions, as presented in Sect. 1.2. Formal methods can also be developed for modeling, requirements specification, design, and validation of safety-critical systems.

Railway projects are governed by texts and standards which aimed to define and reach objectives of Reliability, Availability, Maintainability, and Safety (RAMS). The three standards CENELEC EN 50126, EN 50128 and EN 50129 can cover aspects related to the system's safety, comprising material and/or software elements. The CENELEC standards are an adaptation of the IEC 61508 generic standard to the specific railway domain. The standards can be applied to urban railway applications, such as mass transit, and classical railway applications, such as HST, conventional trains and freight [17]. The organization and application domain of the standards are presented in Fig. 1.2:

- The EN 50126 applies on the complete railway systems by specifying and demonstrating RAMS;
- The EN 50128 and EN 50129 are dedicated to demonstrate the safety of software and hardware respectively, for the signaling subsystem;
- The EN 50159-1 and -2 are linked to the transmission aspects, for open and closed networks respectively.

Finally, it is important to introduce the concept of average Probability of Failure on Demand (PFD) to characterize the different SIL, as presented in Table 1.2. The average PFD is the probability of unavailability of the safety function, leading to dangerous consequences.

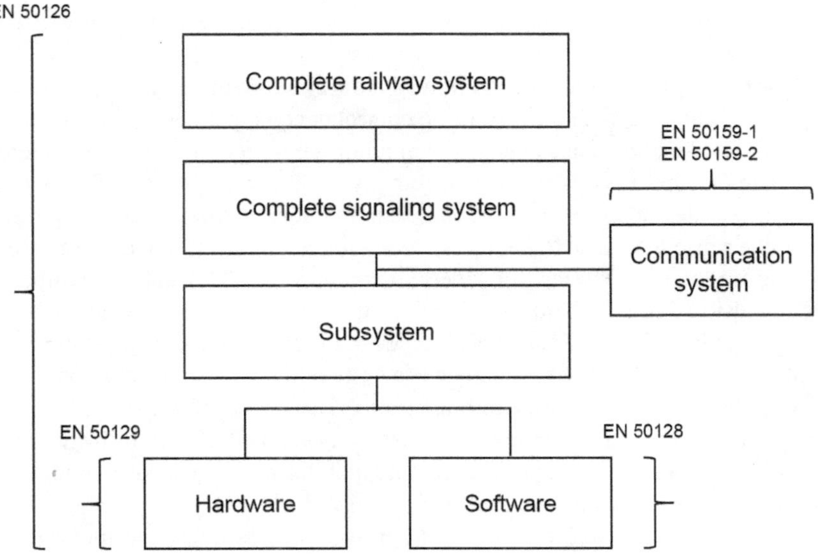

Fig. 1.2 Organization of CENELEC safety standards

Table 1.2 Probability of failure on demand depending on safety integrity level from IEC 61508 stantard

Safety integrity level	Average PFD (PFDavg)
SIL 4	$10^{-5} \leq PFDavg \leq 10^{-4}$
SIL 3	$10^{-4} \leq PFDavg \leq 10^{-3}$
SIL 2	$10^{-3} \leq PFDavg \leq 10^{-2}$
SIL 1	$10^{-2} \leq PFDavg \leq 10^{-1}$

1.1.4 Classifications

Needs for wireless transmissions in the railway environment are quite growing since several years. But it is important to distinguish the different needs depending on intended applications, required throughputs and railway sectors.

1.1.4.1 The Wireless Communication Needs Depending on Intended Applications

Two main kinds of applications can be identified:

- The operational services, that include:
  - The safety-related applications for control and command of trains and signaling, also called "operation control system", aiming to control and monitor the traffic: it includes the control and command of automations, the remote control of trains, the obstacle detection, the driving assistance or the reversal of the trains

for metros without driver [3]. It is a safety application that must meet security requirements;
  - The robust operational voice communication systems: these systems allow trains to communicate with the rail traffic control centers, but also trains drivers, rail traffic and maintenance agents to communicate with each others in conference mode (group calls). It can be noted that some National Safety Administration (e.g. Sweden) also characterize voice communication systems as safety-related, in the sense their unavailability is considered an unacceptable risk for safety;
  - The image transmission for video surveillance (CCTV), the TV-platform or Télévision Semi-Embarquée (TVSE) and the Public Announcement (PA): the first one is a system of cameras disposed in the trains and image transmission to the control center. The second one is a system of cameras disposed on the platform and image transmission to the train driver. TV-platform is used for metro applications, TVSE systems are currently developed for the railway applications. All these systems perform remote monitoring to ensure the conditions of safety and exist mainly for metro applications;
  - The maintenance and diagnosis: these two systems allow the monitoring of infrastructure and rolling-stock. Several systems can be developed, such as track monitoring systems and health monitoring systems;
  - The information to passengers: it consists of information transmission from the ground to passengers in trains, such as train related environment information and train schedule.

- The services to passengers: it comprises mainly the Internet access on board trains, but also infotainment (information + entertainment), Video on Demand (VoD), video streaming and Passenger Information System (PIS).

These applications have access to appropriate services and traffic classes and can even share a same medium. CCTV was initially classified as a non-critical application, that is related to embedded surveillance. However, video is also used more recently for remote control of trains or for other safety-related applications for control of trains, such as reversal of metro trains, and driverless metro.

1.1.4.2 The Wireless Communication Needs Depending on Throughputs

The applications can also be divided depending on the required throughput to ensure the services:

- The applications requiring high throughput: all the services to passenger, such as Internet on board, but also all applications relying on image transmission, such as CCTV and TV-platform, some systems for maintenance applications, but also CCTV used for control of trains;

Table 1.3 Average throughputs depending on application for the metro case

Application	Average throughput
Control command	100 kbps (downlink/uplink)
Operational voice	150 kbps (downlink/uplink)
CCTV	50 kbps (downlink); 2–6 Mbps per train (uplink)
TV platform	2 Mbps per train (downlink); 20 kbps (uplink)
Maintenance	200–500 kbps (downlink)
Passenger information	100 kbps (downlink)

- The applications less consuming in terms of throughput: it includes all the data related to the control and command and signaling system, the operational voice communication systems and the passenger informations.

Thus, applications not involving security require globally much throughput. The needs in terms of throughput depend on the transport operator requirements and can go up to several hundreds of Mbps, the required packet error rate is in the order of 10^{-3}. Telecommunication applications that involve safety are demanding in terms of robustness and availability but the amount of information exchanged can be low. This trend tends to evolve recently, with the introduction of image transmission for control of trains and maintenance, as it will be presented in Chap. 2.

Table 1.3 summarizes an example of the average current needed throughput for the specific case of metro [source Alstom].

The railway wireless applications can be described in terms of some specific characteristics, such as amount and type of data to be transmitted on each link (uplink/downlink) but also in terms of some Key Performance Indicators (KPI), such as end-to-end transmission delay, transmission periodicity, packet loss or bit error rates. QoS can be measured for instance in terms of transmitted packet error rates. For control and monitoring applications, transmitted packet error rates must not exceed 10^{-3} value for which different retransmission protocols are in place to ensure the level of transmission security in terms of railway transport.

1.1.4.3 The Wireless Communication Needs Depending on Railway Sectors

A last classification can be performed regarding the railway sectors, that has a quite important impact on wireless needs. We can then distinguish:

- The metro applications;
- The tramway applications;
- The conventional trains applications;
- The HST context.

Depending on railway sectors, business model are quite different and requirements for the different wireless applications can be quite changing. For the metro applications, the underground environment is conducive to the deployment of broadband wireless communication systems, mainly based on the Wireless Fidelity (Wi-Fi) like standard. Many wireless transmissions can be performed for Communication Based Train Control (CBTC) and other applications requiring large bandwidth, such as CCTV and TV platform.

For the tramway applications, it is quite different from metro as from a signaling point of view, it does not include CBTC (even if some exceptions exist) and the way of driving the tram is closer to the bus. From a telecom standpoint, this means that the availability requirement is not as strong as for metro, so some telecom link could be supported by public telecom operator.

For the railway domain (conventional trains and HST), contrary to urban, where the current Wi-Fi technology can handle both CBTC and also non-critical services like PA, PIS, CCTV or TV platform, GSM-R is a narrow band communication system that can only bear critical services like signaling and operational voice. No broadband link can then be brought easily for train-to-ground communications. Few CCTV systems are deployed in trains and bringing Internet on board is a quite difficult task, as it will be presented in the Chap. 2. Furthermore, in the context of HST, one of the main concerns for the use of wireless technologies is the consequence of high speed on the effective technologies performance, particularly concerning throughput, mobility management and packet losses.

Thus, needs in terms of capacity and throughput depend on the targeted applications. It is important to realize an inventory of the railway user needs in terms of wireless communications between trains and wayside for different kinds of applications. The railway needs include external needs for commercial services to train passengers, and internal needs for operators for operational services. Several parameters can define the railway needs in terms of wireless communications:

- The railway network which has a direct impact on ground infrastructure for wireless communications between trains and wayside;
- The train performances: currently in France, HST run at 300 km/h but should grow up in the next years until 350 km/h. The impact of the velocity has to be considered for handovers management and Doppler effects on signals for instance;
- The train characteristics: the capacity of the trains or the specific design (such as two-levels trains in France);
- The type of application (commercial or operational);
- The type of user for services to passengers (business user or common user).

1.1.5 Basic Architectures for Train-to-Ground Wireless Communications

Railway domain requires specific needs for functional and operational applications, such as train control and command and other information systems, compared to the public telecommunication sector.

On the one hand, railway domain suffers of stringent features, such as high mobility, high handover rate, stringent QoS indicators, reliability. In addition to all these parameters, signaling systems are demanding in communication availability because any communication loss leads to a disruption of the signaling system with an emergency brake of the train. The embedded information in these systems is based on very strict safety rules.

On the other hand, favorable characteristics can be highlighted. For instance, trains are guided transport, constrained to move on rails and to perform a previously known route. The mobility pattern and the data traffic profile can be predicted. It is then possible to identify and predict the "worst cases" to be supported by the network architecture [18–20]. From these observations, it is obvious that a specific communication architecture for railway applications is needed [21–23]. All the train-to-ground services needs, that will define the global architecture, will be detailed in the Sects. 1.2 and 1.3 of this chapter.

Many works performed on connectivity on board trains showed that a unique access terminal on the train, also called "mobile router", represents the best technical solution to optimize performance and throughput, the Train Access Terminal (TAT) [7, 24–28]. This terminal can support one or several technology types. Antenna mounted on the outside of the train connects the TAT. The incoming signal from the TAT then fed access points in the carriages in train. We talk about a two-hop communication solution.

The benefits of using such an architecture is that the TAT can manage different types of technology, and perform some smart agility to select the best means of communication and then to rise the obtained throughputs. For the services aiming at connecting passengers to Internet, such an architecture brings various advantages:

- Avoiding Faraday cage and re-enforcing signal power and sensibility with the roof antenna;
- Reducing the handover of all passengers' connections into a single terminal handover, which aggregates all the traffic. Indeed, using a one-hop structure, that is to say each passenger connecting via its own terminal, leads to a huge handover burden, several handovers (of each passenger) having in fact to be performed simultaneously, which consumes a lot of radio resources;
- Abling to provide the sum of aggregated signals from various Internet Service Providers (ISP) versus the connection to only one single ISP by the passenger device;
- Abling to provide Wi-Fi connection inside the train instead of 4G connection, assuming that all passenger devices are not equipped with a 4G subscription, but all equipped with Wi-Fi.

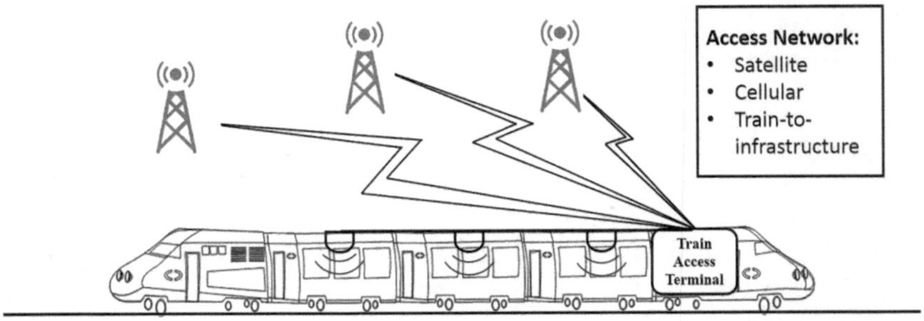

Fig. 1.3 Illustration of the train access terminal

However, it is important to notice that such an architecture can imply interferences between mobile communications inside the train and the external communication. The basic architecture relying on the TAT is illustrated in Fig. 1.3.

Many researches that will be presented in the Chap. 2 stated that the TAT approach works better by combining a set of several wireless technologies instead of only one way to provide broadband communication [29, 30]. There is one technology used to provide connectivity. A second technology, so-called "gap-filler", is used when the first one is no more available. Such a mechanism implies a preferred technology to provide the connection [7]. Criteria to determine on which technology the terminal has to switch are the availability of the links, the quality of the connection (signal strength), delays, throughput and/or economic considerations (costs). Several works and studies were performed on the design of such an architecture, they are described below.

A general Train-to-Ground Network Reference Model (TGNRM) is presented in [21]. The basic architecture relies on three main parts: the radio access network, the aggregation network and the traffic control center. The architecture relies also on the TAT concept. It is important to highlight that this reference architecture can represent a base for any train-to-ground system, except the architecture relying on satellite technology and High Altitude Platforms (HAPs).

The radio access network represents the technology used to connect the train to the ground. The different radio access technologies, such as cellular networks, Leaky Coaxial Cables (LCX) or Radio-over-Fiver (RoF), will be presented in details in the Chap. 2. The aggregation network aggregates the data traffic coming from the access network via the Base Stations (BSs) deployed along the track. The aggregation can rely on different technologies, such as Integrated Services Digital Network (ISDN) (used for the ERTMS system for instance), Ethernet over Fiber, etc. The main requirements of the aggregation network are:

- High Data Rate support;
- Wide radio coverage to allow cost effective deployment and optimize handovers;
- End-to-end QoS support;
- Low latency;
- Support for advanced security scheme;
- ...

The Intelligent COMmunication (ICOM) architecture was developed in the framework of the INTEGRAIL project [22, 23]. The approach of the project was to integrate several wireless railway information systems, dealing with operation, traffic management, rolling-stock and infrastructure domains. The project defined then an ITS architecture based on middleware and IP-based solutions, in the railway domain. The project aimed to create a holistic and coherent information system in order to obtain higher levels of performance in terms of capacity, average speed and punctuality in addition to an optimization of the resources usage [22]. The project proposed new procedures to enhance telecommunication opportunities in the case of multi-user scenarios, such as railway undertakings, railway customers and other railway entities [31]. The ICOM architecture defines a reference model for interoperability standardization, as seen in Sect. 1.1.2.

The solutions for an Internet access on board trains, that will be presented in Chap. 2, rely mostly on a specific basic architecture, depending on the technology used, developed mainly as part of a research project.

For solutions relying on satellite technology, the MOWGLY project proposed an architectural model for implementing wireless broadband access on board moving vehicle with QoS similar to the traditional terrestrial networks [31, 32]. The MOWGLY architecture is based on the use of the Eutelsat GEO satellite constellation, operating in Ku band. The system relies on Digital Video Broadcast Standard 2 (DVB-S2) and Digital Video Broadcast Reverse Channel Satellite (DVB-RCS) standards at physical layer. The solution integrates satellite and terrestrial links in a hybrid system that takes advantages of both technologies, as explained in the previous section.

The MOWGLY architecture then relies on the principle of a "multipath-routing" technology able to select the most appropriate technology and switch the data flow from a link to another. Thus, the final QoS perceived by the end user depends directly on the routing technology. The MOWGLY architecture is divided into 3 segments, as presented in Appendix B.4. The Demand Assigned Multiple Access (DAMA) is used for the reverse link. On the downlink, DVB-RCS based on Multi Frequency Time Division Multiple Access (MF-TDMA), Quadrature Phase Shift Keying (QPSK) modulation scheme and Turbo-Coding is used.

For train-to-infrastructure solutions, the FAMOUS architecture [33] and the WIGWAM research project [34] were proposed, both in 2005.

The FAMOUS architecture was designed to support broadband Internet access for FAst MOving USers. The works were performed by a team of researchers in Belgium [29, 33, 35–40]. The overall FAMOUS network architecture is described in [33]. The objective is to perform the data transport between the fast moving users and the service provider. The architecture consists of two parts:

1. the access network: it represents the wireless network with base stations along the train tracks and provides the last hop communications for the TAT;
2. the aggregation network: it performs the link between the access network and the service providers networks, such as telephone operators or Internet Service Providers (ISPs).

Broadband mobile networks require the use of small cells, leading to a dense placement of antennas along the trajectory of the train. The train connects then the closest wireless base station while passengers are connected to the internal network on board. The base stations are grouped in access networks and when the train is moving, the train's wireless connection hop from one antenna to another, while the passengers keep a seamless connection inside the train. The micromobility protocol allows to keep a seamless connection between the passengers and the aggregation network.

Furthermore, the mobility of the passengers is not limited to a single access network. The connection with the mobile passenger hop from an access network to another. The connection point between the access network and the aggregation network, so-called Access Gateway (AGW), then changes too. The aggregation network is then in charged of data traffic transport from the AGWs to the the service providers. The connection between aggregation networks and service providers is performed by the Service Gateways (SGWs). The mobility between the SGWs and the AGWs is managed by a macromobility protocol at aggregation network level. The issues concerning the aggregation network are discussed in [35, 36, 39]. Two scenarios are considered: the car scenario and the train scenario. For the latter, the system relies on RoF and moveable cells, in order to reduce the handoff times.

The WIreless Gigabit With Advanced Multimedia (WIGWAM) research project aimed to design a 1 Gbps system concept for the home/office, public access and high velocity scenarios (broadband access for trains and highways). The cross-layer project covers the physical layer (PHY layer) up to the protocol layer, and integration into existing infrastructure [34].

At PHY layer, Orthogonal Frequency Division Multiplexing (OFDM) is used as modulation technique. Multiple antenna at transmission and reception can be used for spatial diversity. The presence or not of Channel State Information (CSI) at transmission allows adapting the flexible transmission scheme. Possible spatial equalizers available are a linear processing, a successive interference cancellation and sphere decoding. Possible coding schemes that can be used are the Low Density Parity Check (LDPC)-Codes or the multi-level coding. Other design parameters are settled, such as the guard interval, the subcarrier spacing, the Fast Fourier Transform (FFT) size and the symbol length. All these parameters are tuned by measurements and simulations to increase data rate and system efficiency.

Regarding MAC layer and mobility support, a centralized Orthogonal Frequency Division Multiplexing Access (OFDMA) based solution is required and the main objective is to provide seamless connectivity by handover mechanisms [34]. The WIGWAM works were pursued by the deployment of the system in the Shanghai Transrapid, which will be presented in the section dedicated to RoF systems in Chap. 2.

For the WiMAX technology, [21] proposed the WEWBRA architecture, which is a WiMAX based extension of a Metro area Ethernet network to a Wide Broadband wireless area Network for train-to-ground communication in the Railway scenario, in addition to a IEEE 802.16 Reliable Mobility Pattern Aware (RMPA) handover scheme for the railway domain. Finally, in [26], the Seamless Wireless Internet for

Fast Trains (SWiFT) architecture is introduced. It is based on the deployment of WiMAX (IEEE 802.16m) base stations at the trackside and an optical backbone to link these base stations and the global Internet. The proposed architecture is used in conjunction with the IEEE 802.21 standard, that supports algorithm enabling seamless handover between networks of the same or of different types.

In the French research project CORRIDOR [19, 20], dealing with Cognitive Radio for railway applications, the overall architecture of a communication system was designed, making the best use of Cognitive Radio. The proposed solution is a cross-layer architecture for interacting with heterogeneous radio access technologies. A hierarchical database system allows learning in a local database and the consolidation in a central data base. The Cognitive Manager uses this information to control all components at all levels of the stack (software, operating system, hardware).

Given all these works, we propose a generic basic architecture for train-to-ground communication illustrated in Fig. 1.4. The architecture consists of (1) the Radio Access Network, (2) the Aggregation Network, (3) the Service Provider Network. The only case that does not fit this architecture is the satellite network, which is directly linked to the train and to the traffic control center. In this case, the satellite radio link represents a "data pipe" that could be considered by a mobile aggregation layer (e.g. selection of the best radio link on-board).

Table 1.4 summarizes the different train-to-ground architectures presented in this part, in terms of chosen radio access network and aggregation network.

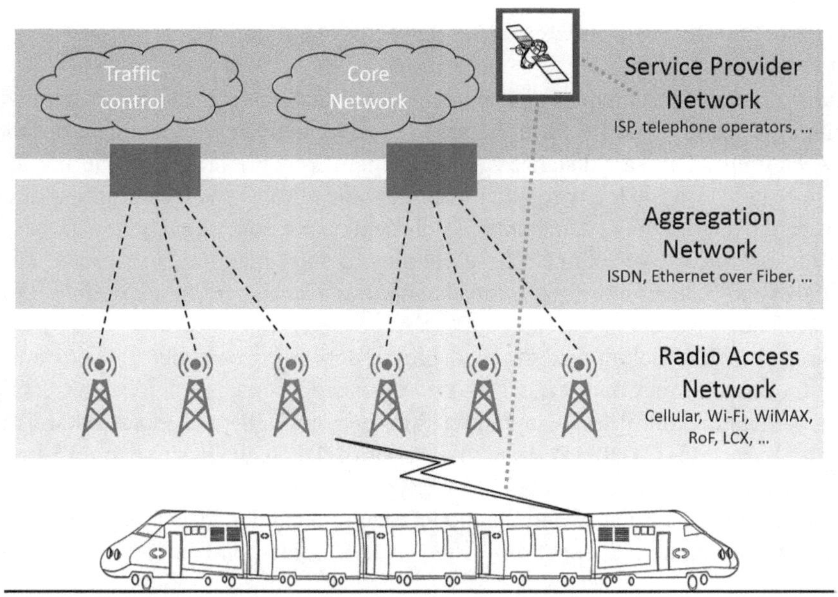

Fig. 1.4 Illustration of basic architecture for train-to-ground communications

Table 1.4 Comparison of train-to-ground architectures

	Radio access network	Aggregation network
ERTMS	GSM-R	ISDN
ICOM	Heterogeneous	Heterogeneous
FAMOUS	RoF	Ethernet over Fiber
SWIFT	IEEE 802.16m	Ethernet over Fiber

1.2 The Needs for Operational Applications

The operational applications relying on wireless communications can be divided into three main categories:

- The communication and signaling systems for train control and command;
- The monitoring systems to limit the risk of injury to persons and damage to property and ensure safe and reliable operations;
- The video applications, such as CCTV and TV platform.

1.2.1 The Communication and Signaling Systems

1.2.1.1 The Different Strategies for Signaling Applications

A large number of techniques and strategies for signaling applications are deployed around the world. They all have the same basic objective: to keep a safe distance between trains. This safe distance can be maintained by measuring the current train position, its relative velocity to other trains and the other trains locations and directions of movement in the same area. All these data are continuously transmitted to other trains via wireless links. These continuous informations about trains' close area allow to reduce inter-train intervals and thus increase traffic capacity without infrastructure investments.

Several such signaling systems are deployed depending on country: the Communication Based Train Control (CBTC), the Advanced Train Control System (ATCS), the Command, Control and Communication System (C3S), the Incremental Train Control System (ITCS), the Positive Train Control (PTC), the Positive Train Separation (PTS) or the European Train Control System (ETCS). The standard IEEE 1474 establishes the performance and functional requirements for a CBTC system [41].

The main operational functions, often linked to each others, can be identified below:

- The Automatic Train Protection (ATP): ATP is a general function, that consists of a railway technical installation to ensure safe operation in the event of human failure. Different systems are deployed around the world relying on inductive systems, cab

signaling or radio-based systems. The communications between the wayside and the train is then assured by inductive loop and radio frequency transmissions;
- The Automatic Train Control (ATC): ATC is a general class of ATP that involves a speed control mechanism in response to external inputs. It is considered to be the safety-critical part;
- The Automatic Train Operation (ATO): ATO is an operational safety enhancement device used to help automate operations of trains;
- The Automatic Train Supervision (ATS): ATS refers to a system within an ATC system which monitors the system status and provides the appropriate controls to direct the operation of trains in order to maintain intended traffic patterns and minimize the effect of train delays on the operating schedule.

A CBTC or an ETCS system is described as an ATC system using high resolution train location determination; continuous, high-capacity, bidirectional train-to-wayside data communications; and train-borne and wayside processors capable of implementing ATP functions, and optional ATO and ATS functions [41]. CBTC applications include the train localization, the train-to-ground transmission and all the communications between the computers in the train and the ground computers [3].

Safety-critical applications generally rely on low throughput (10–100 kbps) but require a high availability (at least 99.999 % of the time) and high levels of robustness and reliability (typically a packet error rates of 10^{-3} for 200 ko length packets) [3]. Furthermore, KPI have to be checked regardless of the mobility conditions: handover times below 10 ms, time of communication establishment, etc. All these KPI are described in the IEEE 1474 standard [41].

1.2.1.2 The Communication Systems

Punctual or Balise-Based Communication Systems

These systems using balises rely on the transmission of information between two points close to each other. These short range communications are locally valid only in the transmission area of the balise. Such systems operate at frequency range from few kHz to several GHz for applications, such as vehicle identification, localization, integrity control or embedded control automation. For instance, ERTMS/ETCS system (level 1) relies on Eurobalise operating at 27 MHz.

Track circuits are electrical circuits obtained by isolating a section of track known as “Block”, used to detect the presence of a train, transmit side signals on trains as well as maximum authorized speed instructions, the next speed limit and the distance to this limitation.

Continuous Communication Systems Using Magnetic Coupling

Although gradually replaced by radio, these systems are still widely used. The core technology consists of a two-wire line. The electric current flowing in the wires induces a magnetic field which allows the continuous transmission of information detected by a magnetic antenna on board the train. They are economical but are subject to limitations, such as low useful bandwidth and poor Signal-to-Noise Ratio (SNR). The wires can be crossed regularly or not. These systems are widely used for automatic metros.

Professional Mobile Radio (PMR) Applications

The deployment of radio systems for railway applications has increased with the arrival of Professional Mobile Radio (PMR) systems, such as the Terrestrial Trunked Radio (TETRA). PMR was developed for business users who need to keep in contact over relatively short distances with a central base station/dispatcher. PMR is also widely used by emergency services. PMR networks consist of one or more base stations and a number of mobile terminals. Such a system serves a closed user group and is normally owned and operated by the same organization as its users. TETRA systems are based on trunking techniques, that rely on a sharing of the resources of the communications network, thus providing both flexibility and economy in the allocation of network resources. Typically, a communication channel is allocated for the duration of a call and then automatically released to allow it to be used for another call, perhaps between different users on the same system. The technique also enables multiple base stations to be connected and to provide coverage across a wider area than with a single base station [42]. For the urban rail domain, TETRA is used for the operational voice systems in the metro/tram, as well as for the data transfer of the Automatic Vehicle Location System (AVLS) for Tramway.

Other communication systems can also be used in the railway context, such as the GSM-Railway (GSM-R) system (GSM—Global System for Mobile communications, technology customized for train-to-ground communications) or the Enhanced Position and Location Reporting System (EPLRS). While the TETRA system is mainly used in metro and for emergency services, GSM-R is used for mainline railway applications, it will be presented in the following.

Wireless Local Area Networks (WLAN)

The emergence of the Wireless Local Area Networks (WLAN), such as Wi-Fi IEEE 802.11a/b/g/n, leads railway operators to rely on the potential of these inexpensive systems for control and command applications. Specific protocol layers for mobility and safety management in addition to a suitable radio engineering have to be implemented to ensure safety and security of operations. These systems are mainly deployed for metro applications (New York, Marmaray, Beijing, Shanghai) [3].

Finally, we can focus on the IEEE 802.11p standard which is an approved amendment to the IEEE 802.11 standard to add Wireless Access in Vehicular Environments (WAVE) [43]. It defines enhancements to 802.11 required to support ITS applications, including data exchange between high-speed vehicles and infrastructure in the licensed ITS band of 5.9 GHz (5.85-5.925 GHz). The IEEE 802.11p relies on a fast adaptation to rapid changes occurring in a highly mobile vehicular network, specifying a set of parameters for the the handoff process.

The IEEE802.11p has been experimented in urban railway transports, for CBTC applications [44] or Passenger Information System (PIS) for instance [45]. However, a recent paper [46] highlights that WAVE is largely developed for vehicle applications but few for applications in the railway domain. In this context, safety has to be considered with highest priority. Specific protocol architecture and technical considerations are presented in the paper to use IEEE 802.11p standard in the railway context.

The initiative to make the use of 5.915-5.935 GHz official for CBTC was pushed by UITP, RATP, STIB, ALSTOM, SIEMENS, and relayed by European Telecommunications Standards Institute (ETSI) to the European European Conference of Postal and Telecommunications Administrations (CEPT). This is conflicting with the request of ITS for the same band. Moreover, the study of using ITS protocol through 802.11p for CBTC does not seems to be fruitful.

1.2.1.3 The ERTMS System

The European Rail Traffic Management System (ERTMS) is composed of the ETCS signaling system and the GSM-R communication system. ERTMS was introduced in 1994 by the European Union [47]. The aim of the ERTMS system is to answer the needs of operational safety, critical-safety measure, reduction of the cost of signaling and communication systems, improvement of the conditions of construction and operation of railways, particularly in terms of international traffic, harmonization of signaling systems and cross-border interoperability throughout the European Rail Network. ERTMS aims to create a single Europe-wide standard for railway signaling with the final aim of improving the competitiveness of the rail sector. Two working groups were set up by the European Commission. The first one was in charge of investigating a new communication system and introduced the GSM-R system for internal voice and data communications. The second one defined a new signaling standard and introduced the ETCS system. ERTMS addresses the management of rail traffic, in terms of efficiency and effectiveness. The main characteristics are [2]:

- The control and command of trains to ensure safe operation;
- Standardized signaling interfaces to enable unrestricted travels across borders;
- Traffic management systems to optimize capacity of the railway lines.

ETCS

ETCS is the signaling element of the ERTMS system, that may be smarter, safer and usable worldwide. It also provides functionality to allow advanced supervision of rail track equipment and rolling-stock. It manages an advanced ATP, the interface to interlocking and cab signaling technologies. ETCS aims to simplify train driving but also reducing costs of investments and maintenance of fixed installations, increasing the traffic capacity and improving the average speed of trains.

Three levels of ETCS are defined. The levels are defined from the way the track is equipped and the way the information are transmitted to the train:

- Level 1: it represents a cab signaling that can be used alone or in conjunction with existing signaling systems. Lineside signals are generally retained and a block

control is achieved in the conventional manner by the interlocking, based on the information detected by track circuits or axle counters. The system is based on punctual track-to-train communications. Balises transmit route data to trains and an on board computer constantly monitors and computes the maximum allowed speed and the braking curve;
- Level 2: at level 2, a Radio Block Center (RBC) transmits continuous data to trains, using the GSM-R technology. The lineside signals are used as a backup. Safer operation at higher speeds are then possible thanks to a near instantaneous communication between the train and the RBC. The balises are used here as passive positioning beacons. The information on position and direction of each train is automatically reported to the RBC at regular intervals. The track detection still remain in place at the trackside;
- Level 3: the level 3 is not defined yet. However, the objective at this level is that ETCS implements a fully continuous radio-based train spacing. ETCS replaces the lineside signals and the trackside detection devices. All speed and signaling information are based on cab signaling. The location of the train is determined by the train odometry and reported to the RBC via the GSM-R link. Thus, interlocking no longer controls train spacing and capacity is optimized. A great simplification is brought at this level with cost reduction of equipment at trackside and a flexible structure no more based on fixed block intervals, we talk about "moving block". In addition, the information on train integrity is also transmitted from the train to the control center.

GSM-R

GSM-R is a wireless communication standard based on the GSM public telecommunication standard (phase 2+) [47]. It was specifically developed for railway communications applications. It was adopted by the UIC because it was the only technology in commercial operation with a great potential to fulfill requirements for the railway services. Some features had to be still added to the standard. It allows:

- For trains, to communicate with railway traffic control points;
- For train drivers, traffic and maintenance agents to communicate with each other in a conference mode (group calls).

It also allows the transmission of ETCS data. The EIRENE standard (European Integrated Railway Radio Enhanced NEtwork project)—MORANE (Mobile Radio for Railways Networks in Europe) specified functional use of communication without failure up to 500 km/h relying on GSM-R, such as:

- Communications between regulator and driver or with agents;
- Point to point communications or group calls;
- Priority calls level;
- Automatic Train Control;
- Remote train control;
- Railway emergency call;
- Track maintenance;

- Passenger services.

Thus, GSM-R allows digital transfer for replacing all wired communication systems and analog railway radio networks existing in each country that are incompatible with each other. GSM-R is a platform for voice and data communications between the various members of the railway team: drivers, regulators and members of the operating team. It provides features such as Voice Group Call Service (VGCS), Voice Broadcast Service (VBS), localized calls and calls preempting in case of emergency and lack of resource. This standard completes interoperability through the use of a single communications platform.

GSM-R allows, in case of ETCS level 2 and 3, transportation of railway signaling information directly to the driver, thus facilitating a higher speed train running and then an optimized traffic capacity, while maintaining a high level of security. Specifications finalized in 2000, are based on the European MORANE project. The specifications are maintained by the ERTMS UIC project together with ERA (in charge of interoperability). GSM-R was chosen by 38 countries worldwide, including all member states of the European Union, who have the legal obligation to install it in the new or updated railway lines, as well as Asia, Eurasia and North Africa. GSM-R generally uses dedicated Base Transceiver Stations (BTS), close to the railway. The distance between two BTS is seven to ten kilometers. This proximity creates a high degree of redundancy and greater coverage and reliability. The train continuously maintains a digital modem connection to the train control center. This connection has a higher priority to other users. If the modem connection is lost, the train stops automatically, in case of ETCS Level 2, and this would occur after a timer expiration which allows for the communication to be re-established if possible. In Germany, Italy and France, the GSM-R network consists of around 3,000 or 4,000 BTS.

In Europe, the GSM-R uses the dedicated following frequency bands:

- 876–880 MHz: for transmitting data (uplink);
- 921–925 MHz: for receiving data (downlink).

GSM-R is also operated in 1800 MHz band, especially in Australia and some African countries.

The spacing in frequency between two physical channels is 200 kHz. The modulation type is the Gaussian Minimum Shift Keying (GMSK). GSM-R relies on Time-Division Multiple Access (TDMA), meaning that the time division multiplex transmission of data are organized for each carrier (or physical channels) by periodic TDMA frame (4.615 ms period). Each TDMA frame is divided into 8 time slots also known as logical channels (with a duration of 577 microseconds each) composed of 148 bits of information. GSM-R uses a lower extension 900 MHz GSM frequencies.

Unlike services such as on board Internet access and video surveillance, the most stringent requirements for safety-related communications are more related to the service reliability and availability rather than data rate transmission. Such service should be constant, uninterrupted with high availability. Since a safety-related service is more important for railway operation than, for example, on board Internet access, priority mechanisms are necessary. This can be implemented in many ways. However, the system can use a different wireless channel for different services, or even use higher QoS standard for critical information than for non-critical information. No matter what approach is used, the supported QoS solutions should guarantee non-critical services don't penalize critical services. QoS on GSM-R system is high due to security level of ERTMS for train traffic (railway signaling functions).

Currently, works are being performed at the European Railway Agency (ERA), with the involvement of UIC and ETSI and support of GSM-R Industry Group, in order to define the next generation of operational communication system that will replace the GSM-R system. UIC takes care of User Requirements in general, ERA takes care of interoperability considerations at EU level (including spectrum), and ETSI shall be the vehicle for standardisation.

The GSM-R continuity is ensured until 2030 but the time to standardize a communication technology for mission-critical applications such as railways is very important [48]. Indeed, the main defects of GSM-R are [48]:

- the capacity limitations;
- the high costs (associated to a niche market);
- the long development cycle times;
- a sub-optimal use of available bandwidth, which limits the number of trains in a controlled area;
- the forecast obsolescence of the technology (end of life expected for 2030).

Workshops, interviews and discussions are introduced to discuss about the future system. The specifications should be available by 2018, standards applicable by 2020, in parallel to some trials involving operators and industry. First deployments are expected from 2020–2022. Many scenarios are considered, such as keeping the current GSM-R network (no modification), evolution of GSM-R, unmodified commercial networks, modified/enhanced commercial networks, mission critical/professional networks (such as Public Protection Disaster Relief-PPDR) or hybrid networks. All these strategies and visions will be detailed in Chap. 3.

1.2.2 Monitoring Systems

The rail safety management system is responsible for limiting the risk of injury to persons and damage to property, and ensuring safe and reliable operations. Indeed, the rolling-stock is subject to several environmental elements and activities that have a direct effect on persons and properties. Thus, the railway monitoring systems ensure the care and cure of the rolling-stock and then the safety of railway systems. The constant increase of capacity and speeds of trains makes important the knowledge of the state of the vehicle. Wireless Sensor Networks (WSNs) can be used to monitor railway infrastructure and environment. They can even be used for surveillance applications for safety and security purposes. Several papers in the literature deal with these subjects, relying on video for defect detection and railway inspection [49] or automatic detection of railway surface defects, also called "squats" [50].

Furthermore, the development and miniaturization of WSNs, an essential component of the Internet of Things (IoT), offer large possibilities to railway operators and infrastructure managers. These systems allow the rise of large-scale information for everything related to the rolling-stock, the infrastructure or the equipment diagnosis. Data collection can be related to the weather conditions, the infrastructure and rolling-stock aging, the power consumption, and the maintenance and diagnosis.

WSN techniques are then more and more used to monitor the entire railway system and perform the railway system maintenance. The safety of railroad tracks is provided by track monitoring systems. Several systems are already implemented but enhancement are continuously on-going on this subject. The objective is to measure stress, settlement, degradation, stiffness, friction, twist, defects or impacts from climatic changes. Some systems allow continuous monitoring with immediate processing of data, to observe real-time profile of the track for instance [2]. Such systems are composed of sensors, data acquisition system and data processing module. The aim is to increase track lifetime and availability.

On the same principle, health monitoring systems can be implemented to evaluate track condition, prevent damage and perform derailment monitoring. Accelerometer and angular rate sensors are used to represent the degrees of freedom of the car body motion.

Railway Companies developed several systems to perform monitoring of trackside: inspection portal to monitor trains at critical points with the simultaneous acquisition of a 3D profile and a thermal map of the rolling-stock, real-time measurement of wheels' profile using lasers and video cameras, automatic inspection of wheel sets, brake pads measuring systems for predictive maintenance or wheel detector system for control of overheating. Several other applications can be implemented to ensure monitoring of the railway system, such as lightning protection, coach electric safety or traction monitoring. For all these systems, wireless communications are needed to transmit the different informations.

Thus, many monitoring systems are deployed in order to improve maintenance mechanisms. More and more predictive maintenance are required especially to deploy cost efficient and reliable high capacity infrastructure. For this, better maintenance strategies have to be developed especially to perform predictive maintenance based on reliable sensors. Fewer defects and less risk of operational unreliability could then be observed. Intelligent maintenance can include predictive, risk-based or condition-based maintenance and be performed from measurement and monitoring tools to provide static and dynamic data from all relevant components of the rail infrastructure.

Train-to-ground communications can then be used for proactive maintenance applications. These systems can rely on taking images of railway infrastructure by train. Such systems require a large capacity for the uplink in order to upload data from train to ground.

1.2.3 Video Surveillance (CCTV)

Image processing allows ensuring safety and security of railway systems. Thus, if an incident related to security occurs, the monitoring system supports the operator in taking the good decision and provide all required informations. Many projects and systems were studied for security applications. Networks of sensors can then be used for potential incident detection and alert to a control center. Information can be exchanged between the different sensors through a Local Area Network (LAN). Wireless communications between sensors allow the monitoring of larger areas with higher efficiency.

Video surveillance is today a major need for railway operators. Unlike the on board Internet access service, the video surveillance systems require a high capacity on the uplink transmission. Two cases of use can be identified for video surveillance applications:

- View of video records in case of security services (criminal identification, crime evidences);
- Case of doubts in real time ambiguous situations.

Both cases could require wireless transmissions from train to wayside. The first case does not require a high QoS but very high throughputs to download huge volume of data. In the second case, less capacity could be needed, by transmitting for instance images from one camera and by degrading the image quality. To have an idea, a downlink throughput of 512 kbps is sufficient (just to have a control/command on the CCTV application). The uplink throughput is evaluated at about 1 Mbps, considering the transmission of the flow of one camera, with 25 images per second, and a rate of compression of 1:50 (using H264 compression algorithm for instance).

As for multimedia applications that will be seen in the following, CCTV has some QoS criteria, such as:

- Delay;
- Delay variation;
- Information loss: packet error rates of about 10^{-2} for 1 Mo length packets;
- Throughputs: several decades of Mbps;
- Cohabitation with other wireless systems without interference.

All these monitoring systems and WSN techniques require the transmission of data requiring more and more bandwidth. The different existing systems are presented in details in Chap. 2.

1.3 The Needs for Services to Passengers

1.3.1 Introduction to Internet on Board Trains

Current public telecommunication services increased the needs of mobility services. Indeed, wireless communications have been deeply integrated into people's life. Peoples are more and more demanding in terms of transmission rates. High Speed Downlink Packet Access (HSDPA) allows throughputs up to 28 Mbps using 5 MHz of bandwidth. Currently, Long Term Evolution (LTE) is largely deployed all around the world. Theoretical throughputs can reach 300 Mbps on a 100 MHz bandwidth. All these recent evolutions concern also the transportation domain [28].

Many recent technological advances were observed: a miniaturization of devices, an ergonomic use of the devices, the development of wireless communications between the devices, the significant increase of data transfer and the multiplication of functionality of the devices [32].

It is also important to note that railway domain is a more and more competitive environment and it becomes important for railway operators to make travel more comfortable and pleasant by offering new services or by improving existing services relying on telecommunications. The main commercial service is the Internet access on board trains. We then focus on this subject in this part dedicated to services to passengers.

The first solution for providing broadband Internet access on board trains could be that users directly connect their devices (smartphone, laptop, tablet) via their own mobile operator [7, 51]. However, this solution requires appropriate cells management and ability, and sufficient radio coverage along the railway line. Indeed, public networks design does not take always in consideration the railways network. In addition, as already mentioned, the metallic structure of carriages behaves like a Faraday cage, which causes signal attenuation up to 15 dB, and even 30 dB in worst cases [28]. To overcome this problem and improve performance, repeaters may be installed on trains. However, installation and maintenance of these repeaters involve significant costs. Furthermore, performance of these transponders depends closely on the quality of the radio coverage: the amplification of a weak signal causes a low SNR and therefore a "bad quality" amplified signal. Some experiments were performed in 2007 by Sauter [52]. The experiment corresponds to a direct connection of a terminal in the train via HSDPA. The tests were performed on a German Intercity Express (ICE) HST from Paris to Frankfurt. Peak throughputs of 1.5 Mbps and average throughputs of 850 kbps were observed. However, communication failures are often encountered. It is clear that other solutions need to be developed.

Internet access on board trains is not a standardized technology yet. A lot of different solutions and architectures were then developed to bring Internet on board trains [32]. Finally, regarding required bandwidth, if we consider that a train contains typically 1500 passengers, the bandwidth requirements of several Gbps per train is not unrealistic [33].

1.3.2 User Needs

As already mentioned, user needs depend on type of services. For business users, throughputs can be limited. The main Internet applications needs for this type of users are browsing, emailing (with attached files), which does not require real-time process. Common user has a complete different behavior than business one. Common user requires large bandwidth and often real-time web applications (games, video, chatting, etc.).

Internet services require especially throughputs on the downlink and less throughputs on the uplink. For business users, throughputs are 4 times more important for downlink than uplink. For common users, this rate is around 9 [53].

In addition to Internet access, multimedia services can also be offered on board trains. These services target especially common users. It can include video streaming, infotainment, VoD, information on travel conditions, Virtual Private Network (VPN), VoIP, live TV, Peer-to-Peer (P2P) or videoconferencing.

Key factors of QoS for multimedia services are given in the ITU-T G.1010 standard [54]. This document aims to determine the parameters that govern end-user satisfaction for applications involving voice, video, image or text.

Criteria for evaluating QoS of multimedia services are presented in the standard:

- Delay: it consists of the time taken to establish a particular service from the initial user request and the time to receive specific information once the service is established;
- Delay variation: it is generally included as a performance parameter since it is very important at the transport layer in packetised data systems due to the inherent variability in arrival times of individual packets. Applications requiring few delay variation implement techniques to remove or reduce delay variations;
- Information loss: Information loss has a very direct effect on the quality of the information finally presented to the user, whether it be voice, image, video or data.

Multimedia services can be classified in several applications.

Audio applications can be divided as follows:

- Conversational voice: requirements in terms of delay are heavily influenced by one-way delay; for information loss, requirements are influenced by the fact that human ear is tolerant to a certain amount of distortion of a speech signal;
- Voice messaging: requirements are similar in terms of information loss but quiet different in terms of delay where there is more tolerance because of the no direct conversation involved;
- Audio streaming: requirements in terms of information loss have to be tighter than classical conversational voice to provide better quality. As for voice messaging, more tolerance is possible in terms of delay.

Video applications can be divided as follows:

- Videophone: it implies a full-duplex system, combining both video and audio. It has then the same delay and loss information requirements as for conversational

Table 1.5 QoS criteria for classification of multimedia applications

	Interactive (delay <<1 s)	Responsive (delay ≈2 s)	Timely (delay ≈10 s)	Non-critical (delay >>10 s)
Error tolerant	Conversational voice and video	Voice/video messaging	Audio and video streaming	Fax
Error intolerant	Interactive games	Web browsing	Messaging, downloads	Background

voice. In addition, a synchronization is necessary between audio and video to provide "lip-synch";

- One-way video: requirements in terms of delay and information loss are similar to audio streaming.

For any data transfer, the prime requirement is to guarantee zero loss of information. The various applications differ in their needs in terms of delay. We can quote for instance web browsing, e-mailing or interactive games.

Finally, among multimedia services, background applications can be found. It consists of applications for which information have to be delivered to the user essentially error free. Delay constraints still remain because of the uselessness of too late received data. Among these applications, there are for instance fax applications or Short Message Service (SMS).

The different applications can then be classified depending on the QoS criteria. They are presented in Table 1.5.

For instance, conversational voice and data or messaging based on voice or data require low delay but can suffer of information loss. On the contrary, messaging or downloads (File Transfer Protocol (FTP) for instance) require no information loss but can support more important delays.

All these criteria are based on user perception and then are not dependent on application. It is also important to note that applications cited are examples and the list is not exhaustive. Other applications can be added in the model by their similarity.

The given model provides lower and upper boundaries to be perceived as acceptable for the user. If the upper boundary is passed, the service is considered as unsatisfactory. On the other side, if lower boundary is exceeded, application is still perceived as acceptable but it can be a waste of unnecessary network resources.

The ITU document presents also performance targets for audio and video applications. Performance are presented in terms of delay, delay variation, information loss (packet loss ratio) and other criteria such as "lip-synch".

1.4 Summary on Wireless Communications in the Railway Domain

The chapter was dedicated to an inventory of all the train to ground wireless communication needs in the railway domain. User needs for railway operators gather both multimedia services for commercial services and video and data services for operational services. The fast development of broadband communications for public telecommunications, the multiplication and miniaturization of electronic devices, the increase of speed of trains lead to the development of wireless broadband communications for railway applications.

The safety-critical systems used for the train control and command and signaling require high level of availability, reliability and robustness and recently more throughputs due to the use of images for the train control for instance. For commercial services, high capacities are required in order to transmit a large amount of data. Furthermore, Internet access on board trains requires more capacity on the downlink than on the uplink. For some service, such as video (CCTV) or telephony over IP, synchronous transmission is necessary. For other applications like maintenance, an asynchronous transmission is necessary. Furthermore, capacity on the uplink is required to upload data from train to ground. Two-way transmission is also required in most of the cases.

Chapter 2 is devoted to the presentation of the railway applications requiring broadband wireless communications. Chapter 3 is dedicated to the challenges and the future of the railway communications. The current developments, future technologies and scientific barriers are highlighted.

References

1. Gianosvaldo Fadin. ICT Standardisation on Board of Trains, an overview of current and upcoming standardss. In *CENELEC-ACRI Workshop: Information and Communication Technology On Trains*, Prague, Czech Republic, October 2011.
2. G. M. Shafiullah, A. Gyasi-Agyei, and P. Wolfs. Survey of Wireless Communications Applications in the Railway Industry. In *The 2nd International Conference on Wireless Broadband and Ultra Wideband Communications (AusWireless 2007)*, Sydney, Australia, August 2007.
3. Marion BERBINEAU, Mohamed KASSAB, Christophe GRANSART, Martine WAHL, Yann COCHERIL, Emile MASSON, Divitha SEETHARAMDOO, David SANZ, Hassan GHANNOUM, and Oriane GATIN. Le véhicule connecté dans les transports publics: technologies existantes et perspectives. *REE. Revue de l'électricité et de l'électronique*, (4):60–67, 2014.
4. ERRAC. Strategic rail research and innovation agenda - A step change in rail research and innovation. Technical report, ERRAC, November 2014.
5. Shift2Rail Joint Undertaking, Multi-Annual Action Plan. Technical report, November 2015.
6. Bo Ai, Xiang Cheng, Thomas Kurner, Zhang-Dui Zhong, Ke Guan, Rui-Si He, Lei Xiong, David W Matolak, David G Michelson, and Cesar Briso-Rodriguez. Challenges toward wireless communications for high-speed railway. *Intelligent Transportation Systems, IEEE Transactions on*, 15(5):2143–2158, 2014.
7. Daniel T Fokum and Victor S Frost. A survey on methods for broadband internet access on trains. *Communications Surveys & Tutorials, IEEE*, 12(2):171–185, 2010.

8. R. He, Z. Zhong, B. Ai, L. Xiong, and H. Wei. A novel path loss model for high-speed railway via duct scenarios. In *7th WiCOM*, pages 1–4, Beijing, China, 2011.
9. TS Rappaport, JN Murdock, DG Michelson, and R Shapiro. An open-source archiving system. *Vehicular Technology Magazine, IEEE*, 6(2):24–32, 2011.
10. Qiong Wu, David W Matolak, and Indranil Sen. 5-ghz-band vehicle-to-vehicle channels: models for multiple values of channel bandwidth. *Vehicular Technology, IEEE Transactions on*, 59(5):2620–2625, 2010.
11. Kyle N Sivertsen, Anthony Liou, and David G Michelson. Depth and rate of fading on fixed wireless channels between 200 MHz and 2 GHz in suburban macrocell environments. *Antennas and Propagation, IEEE Transactions on*, 58(10):3353–3362, 2010.
12. RS Thomä, M Landmann, and A Richter. Rimaxa maximum likelihood framework for parameter estimation in multidimensional channel sounding. In *Proceedings of the International Symposium on Antennas and Propagation (ISAP04)*, pages 53–56. Citeseer, 2004.
13. http://www.roll2rail.eu/.
14. Martine Wahl. *Survey of railway embedded network solutions. Towards the use of Industrial Ethernet technologies*. Les collections de l INRETS, March 2010. ISBN 2-85782-682-8.
15. Gianosvaldo Fadin and Paolo Umiliacchi. New standard for multimedia applications on board trains enables improved interoperability. In *9th World Congress on Railway Research*, Lille, France, May 2011.
16. J. Smith, S. Russel, and M. Looi. Security as a safety issue in rail communications. In *8th Australian Workshop on Safety Critical Systems and Software (SCS03)*, pages 79–88, Canberra, Australia, 2003.
17. Jean-Louis Boulanger. Maîtrise du SIL et gestion des certificats - Domaine ferroviaire. Technical report, Dossier Techniques de l'ingénieur, 2011.
18. Marina Aguado, Oscar Onandi, Purificacion Agustin, Marivi Higuero, and Eduardo Taquet. WiMAX on rails. *IEEE Vehicular Technology Magazine*, 3(3):47–56, 2008.
19. Kamal Deep Singh, Priyanka Rawat, and Jean-Marie Bonnin. Cognitive radio for vehicular ad hoc networks (CR-VANETs): approaches and challenges. *EURASIP Journal on Wireless Communications and Networking*, 2014(1):1–22, 2014.
20. Jong-Hyouk Lee, Keshav Dev Singh, Jean-Marie Bonnin, and Sangheon Pack. Mobile data offloading: a host-based distributed mobility management approach. *Internet Computing, IEEE*, 18(1):20–29, 2014.
21. Marina Aguado. *Broadband Communication Architecture for Train to Ground Communication Services: the WEWBRA Architecture and the RMPA Handover*. PhD thesis, Faculty of Engineering of Bilbao, June 2009.
22. Jerome Billion and Didier Van den Abeele. ICOM: A Communication Framework for Interoperable European Railways. In *7th International Conference on Intelligent Transportation Systems Telecommunications (ITST)*, pages 1–6, Nice, France, June 2007. IEEE.
23. R. Shingler, G. Fadin, and P. Umiliacchi. From RCM to predictive maintenance: The InteGRail approach. In *4th IET International Conference on Railway Condition Monitoring*, pages 1–5, June 2008.
24. G. Bianchi, N. Blefari-melazzi, E. Grazioni, S. Salsano, and V. Sangregorio. Internet access on fast trains: 802.11-based on-board wireless distribution network alternatives. In *IST Mobile & Wireless Communications Summit*, pages 15–18, Aveiro, Portugal, June 2003.
25. A. Bocquet and C. Gransart. Benefits of multimodel architecture-based development for railway applications. In *World Congress on Railway Research*, Seoul, Korea, May 2008.
26. Ritesh Kumar, Pravin Angolkar, Debabrata Das, and Ramakrishnan Ramalingam. SWiFT: A Novel Architecture for Seamless Wireless Internet for Fast Trains. In *IEEE Vehicular Technology Conference (VTC Spring)*, pages 3011–3015, Singapore, May 2008.
27. Luc Verstrepen, Wout Joseph, Emmeric Tanghe, Jan Van Ooteghem, Bart Lannoo, Mario Pickavet, Luc Martens, and Piet Demeester. Making a well-founded choice of the wireless technology for train-to-wayside data services. In *9th Conference on Telecommunications Internet and Media Techno Economics (CTTE)*, pages 1–7, June 2010.

28. Yiqing Zhou, Zhengang Pan, Jinlong Hu, Jinglin Shi, and Xinwei Mo. Broadband wireless communications on high speed trains. In *Wireless and Optical Communications Conference (WOCC), 2011 20th Annual*, pages 1–6. IEEE, 2011.
29. Daan Pareit, Walter Van Brussel, Wim Torfs, and Peter De Cleyn. QoS-enabled Internet-on-train network architecture: inter-working by MMP-SCTP versus MIP. In *7th International Conference on Intelligent Transportation Systems Telecommunications (ITST)*, pages 1–6, Sophia-Antipolis, France, June 2007.
30. Pablo Rodriguez, Rajv Chakravorty, Julian Chesterfield, Ian Pratt, and Suman Banerjee. MAR: A Commuter Router Infrastructure for the Mobile Internet. In *2nd International Conference on Mobile Systems, applications, and services (MobiSys)*, pages 217–230, June 2004.
31. G. Barbu. E-TRAIN Broadband Communication with Moving Trains - Technology state of the art. Technical report, UIC E-TRAIN project, June 2010.
32. Sajjad Karim, Shoaib Khan, Ahmed Qureshi, and Imran Daud. Turning need into demand for Wi-Fi broadband internet access on trains. *International Journal of Computer Applications*, 53(4), 2012.
33. Filip De Greve, Bart Lannoo, Liesbeth Peters, Tom Van Leeuwen, Frederic Van Quickenborne, Didier Colle, Filip De Turck, Ingrid Moerman, Mario Pickavet, Bart Dhoedt, et al. Famous: A network architecture for delivering multimedia services to fast moving users. *Wireless Personal Communications*, 33(3-4):281–304, 2005.
34. Gerhard Fettweis and Ralf Irmer. WIGWAM: System concept development for 1 Gbit/s air interface. In *14th Wireless World Research Forum (WWRF'05)*, July 2005.
35. Filip De Greve, Frederic Van Quickenborne, Filip De Turck, Ingrid Moerman, and Piet Demeester. Cost-effective ethernet routing schemes for dynamic environments. In *Global Telecommunications Conference, 2005. GLOBECOM'05. IEEE*, volume 2, pages 6–pp. IEEE, 2005.
36. Filip De Greve, Frederic Van Quickenborne, Filip De Turck, Ingrid Moerman, and Piet Demeester. A new carrier grade aggregation network model for delivering broadband services to fast moving users. *International Journal of Communication Systems*, 20(3):335–364, 2007.
37. Bart Jooris, Piet Verhoeve, Frederik Vermeulen, and Ingrid Moerman. Mobile Communication and service continuity in a train scenario. In *Symposium of the IEEE/CVT*, Enschede, Netherlands, November 2005.
38. Bart Lannoo, Didier Colle, Mario Pickavet, and Piet Demeester. Radio-over-fiber-based solution to provide broadband internet access to train passengers. *IEEE Communications Magazine*, 45(2):56–62, 2007.
39. Frederic Van Quickenborne, Filip De Greve, Filip De Turck, and Piet Demeester. Managing ethernet aggregation networks for fast moving users. *Communications Magazine, IEEE*, 44(10):78–85, 2006.
40. Stijn Verstichel, Filip De Turck, Bart Dhoedt, Piet M. Demeester, and Frederik Vermeulen. On the design of a Train Communication Management platform. In *Symposium on Communications and Vehicular Technology*, pages 29–34, Liege, Belgium, 2006. November.
41. IEEE standard method for CBTC performance and functional requirements.
42. http://www.etsi.org/technologies-clusters/technologies/digital-mobile-radio/private-mobile-radio.
43. IEEE Standard for Information technology - Telecommunications and information exchange between systems - Local and metropolitan area networks - specific requirements Part 11: Wireless LAN Medium Access Control (MAC) and Physical Layer (PHY) specifications Amendment 6: Wireless Access in Vehicular Environments.
44. Li Zhu, F Richard Yu, and Bing Ning. A seamless handoff scheme for train-ground communication systems in cbtc. In *Vehicular Technology Conference Fall (VTC 2010-Fall), 2010 IEEE 72nd*, pages 1–5. IEEE, 2010.
45. Li Zhu, F Richard Yu, Bin Ning, and Tao Tang. Cross-layer design for video transmissions in metro passenger information systems. *Vehicular Technology, IEEE Transactions on*, 60(3):1171–1181, 2011.

46. R. Y. Kim. IEEE 802.1p based WAVE Protocol for Railroad Applications. *IOSR Journal of Engineering*, 6(1):16–19, January 2016.
47. Simon F. Ruesche, Jan Steuer, and Klaus Jobmann. A Packet-Switched Approach to a Train Control System. *IEEE Vehicular Technology magazine*, pages 37–48, September 2008.
48. Almudena Diaz Zayas, Cesar A Garcia Perez, and Pedro Merino Gomez. Third-generation partnership project standards: for delivery of critical communications for railways. *Vehicular Technology Magazine, IEEE*, 9(2):58–68, 2014.
49. Hao Feng, Zhiguo Jiang, Fengying Xie, Ping Yang, Jun Shi, and Long Chen. Automatic fastener classification and defect detection in vision-based railway inspection systems. *Instrumentation and Measurement, IEEE Transactions on*, 63(4):877–888, 2014.
50. Maria Molodova, Zili Li, Alfredo Núñez, and Rolf Dollevoet. Automatic detection of squats in railway infrastructure. *Intelligent Transportation Systems, IEEE Transactions on*, 15(5):1980–1990, 2014.
51. Hassan Ghannoum and David Sanz. Internet Onboard: technical analysis. In *5th International Workshop on Communication Technologies for Vehicles, Nets4cars/Nets4trains*, Lille, France, May 2013.
52. http://mobilesociety.typepad.com/mobile_life/2007/07/hsdpa-internet-.html.
53. H. Philippe, H. Ghannoum, N. Haziza, and M. Berbineau. Railway user needs for high speed trains. Technical report, Deliverable D2.1 - CORRIDOR Research project, 2014.
54. ITU-T. Series G: Transmission Systems and Media, Digital Systems and Networks - Quality of Service and performance, End-user multimedia QoS categories ITU-T. Technical report, ITU, 2002.

Chapter 2
Railway Applications Requiring Broadband Wireless Communications

This chapter is dedicated to all the railway applications requiring broadband wireless communications. The works focus especially on the way to provide Internet access on board trains. The different technologies are reviewed, regarding initial research projects, architectures and existing deployed solutions.

Providing Internet access on board trains ensure broadband links between train and ground, which allows railway operators and infrastructure managers to ensure other applications. Thus, operational applications requiring high throughputs, relying on video transmission for instance, such as predictive maintenance and video surveillance, could be also considered. They are presented in the second part of the chapter.

2.1 Broadband Internet Access on Board Trains

Providing an Internet access on board trains became an important objective for railway operators in a more and more competitive domain. Several technologies and strategies can be implemented, relying globally on the same basic architecture presented in Sect. 1.1.5. A selected survey of the studies, projects and solutions deployed all around the world is presented.

Two issues have to be solved:

- How users access Internet on board (from the train access terminal)?
- How to connect the train (via the train access terminal) to the Internet backbone?

É. Masson and M. Berbineau, *Broadband Wireless Communications for Railway Applications*, Studies in Systems, Decision and Control 82,
DOI 10.1007/978-3-319-47202-7_2

2.1.1 How Users Access Internet on Board?

2.1.1.1 Wi-Fi

Deploying a WLAN such as Wi-Fi within the train is the approach chosen unanimously by all deployed solutions [1, 2]. The deployment of a wired Ethernet network could be considered. However, it causes very high installation costs, especially since it requires equipment of all "connected trains" [3]. Rewiring may also be needed every time the train is reconfigured. In addition, it is generally accepted that replication concept of Wi-Fi access points within the train is not only the best technical solution to create "connected trains", but also the ideal client interface [1]. Wi-Fi is a well-known technology with unlicensed bands, easy to roll out and cost effective. The different standards of Wi-Fi IEEE 802.11 are reminded in the Table 2.1.

IEEE 802.11n allows achieving theoretical throughput up to 450 Mbps on each available band (2.4 and 5 GHz). It improves the previous standards: IEEE 802.11a on the 5 GHz band, and IEEE 802.11b and IEEE 802.11g on the 2.4 GHz band by the following enhancements:

- The standard relies on the Multiple Input Multiple Output (MIMO) technology;
- The aggregation of channels allows increasing the bandwidth to 40 MHz, instead of 20 MHz for previous standards.

In [3], two different topologies are studied in order to construct the network on board train using Wi-Fi technology. The wireless coverage in a single carriage is easily achieved via an Access Point (AP) by using IEEE 802.11a. The issue is the wireless distribution network, whose goal is to interconnect the different carriages. The first topology consists in linking the different cars with IEEE 802.11b via external directive antennas. The second topology is based on the assumption that the access network (IEEE 802.11a in each carriage) is also used as a distribution network between the different carriages. The paper concludes that these propositions still have to be tested by real experiments.

Finally, the last Wi-Fi standard developed from 2011 to 2013 was approved in January 2014, the so-called IEEE 802.11ac. It uses the 5 GHz band exclusively. Theoretical throughput of 500 Mbps can be reached and up to 7 Gbps by using multiplexing and MIMO techniques. The main drawback of this new standard is

Table 2.1 Main standards of the Wi-Fi IEEE 802.11

Standard	Standardization	Frequency	Theoretical throughput
IEEE 802.11a	1999	5 GHz	54 Mbps
IEEE 802.11b	1999	2.4 GHz	11 Mbps
IEEE 802.11g	2003	2.4 GHz	54 Mbps
IEEE 802.11n	2009	2.4/5 GHz	450 Mbps
IEEE 802.11p	2010	5.85–5.925 GHz	/

that terminals have to be specifically designed for this new technology. However, equipped terminals can support 802.11n standard (but not a, b and g ones).

Some researches investigate also very recent technologies, such as the new IEEE 802.11ad, or WiGig (at 60 GHz), and the Li-Fi. They are presented in the following.

2.1.1.2 WiGig or IEEE 802.11ad

The WiGig (Wireless Gigabit—also known as 802.11ad) is a new wireless technology operating at the unlicensed 60 GHz band (9 GHz bandwidth from 57 to 66 GHz in Europe) that will able broadband communications and very high throughput up to 7 Gbps [4–6]. It allows high-speed, low latency, and security-protected connectivity between nearby devices. WiGig technology has a limited transmission distance around several decades of meters. Recent advances of using SiGe and CMOS to build inexpensive 60 GHz transceiver components lead to a growing interest to the 60 GHz radio [5].

WiGig was developed by the WiGig Alliance, which was formed to promote the IEEE 802.11ad protocol in May 2009. The Wi-Fi Alliance subsumed the WiGig Alliance in March 2013. WiGig will then extend the Wi-Fi Alliance vision for seamless connectivity and enable new use cases that complement traditional Wi-Fi. Popular use cases for WiGig include cable replacement for popular Input/Output (I/O) and display extensions, wireless docking between devices like laptops and tablets, instant synchronization and backup and simultaneous streaming of multiple ultra-high definition and 4 K videos.

With WiGig technology now under the wing of Wi-Fi Alliance, the forthcoming WiGig CERTIFIED program will ensure devices provide a great user experience, the latest security protections, and multi-vendor interoperability. Many WiGig CERTIFIED products are expected to be Wi-Fi CERTIFIED as well, and products implementing both WiGig and Wi-Fi will include mechanisms to facilitate seamless handover between the two technologies.

WiGig operating in millimeter waves domain, a specific challenge to overcome is the severe path loss from transmitter to receiver [4]. Typically, WiGig systems will suffer a loss of about 21 to 28 dB relative to the IEEE 802.11n (operating at 2.4 and 5 GHz), because of the shorter wavelength at 60 GHz. Thus, the distance between the transmitter and the receiver have to be reduced and the remained loss has to be compensated by increasing the antenna gain. Increasing antenna gain leads to a narrower beamwidth of the antenna, which requires automated antenna pointing or beamforming. This was not an issue for the IEEE 802.11a/b/g/n standards that use omnidirectional antennas.

The PHY and MAC layers specifications of the WiGig [7] provide similar functionality to the IEEE 802.11a/b/g/n standards, incorporating enhanced operations in the 60 GHz band. The WiGig MAC and PHY specifications, version 1.1, includes the following capabilities:

- Data transmission rates up to 7 Gbps are supported, more than ten times faster than the highest 802.11n rate;
- The 802.11 MAC layer is supplemented and extended, it is backward compatible with the IEEE 802.11 standard;
- PHY layer enables low power and high performance WiGig devices, guaranteeing interoperability and communication at gigabit rates;
- Protocol adaptation layers are being developed to support specific system interfaces including data buses for PC peripherals and display interfaces for HDTVs, monitors and projectors;
- Support for beamforming, enabling robust communication at distances beyond 10 m, is implemented. The beams can move within the coverage area through modification of the transmission phase of individual antenna elements, which is called phase array antenna beamforming;
- Advanced security and power management are widely used for WiGig devices.

Beamforming techniques is an integral part of these specifications [5]. Beamforming utilizes multiple antennas to form a beam toward a certain direction to increase the signal strength. This beamforming gain is achieved by transmitting phase-shifted signals from multiple antenna elements, which are added coherently. Beamforming at 60 GHz can be easier performed compared to the 2.4 or 5 GHz bands. Indeed, antenna sizes are reduced and multiple antennas can be packed in a very small area [5]. In [8], an extra codebook is proposed in order to avoid the signal loss introduced at the intersection of two adjacent beams when employing original beamforming codebook of the IEEE 802.11ad standard. It is based on Maximal Ratio Combining. Performed simulations showed a significant decrease of BER by using the new codebook; a decrease of the BER from 5×10^{-4} to 10^{-4} is for example obtained with a codebook using three antenna elements.

A final point that can be addressed on the WiGig technology is that a large recent literature can be found on the development of antennas for WiGig applications at 60 GHz. In [9], 3D printing technology is used to develop innovating lens design and improve the gain of existing 60 GHz antenna solution. A 10 dB improvement is achieved in the budget link. In [10], the authors developed a magneto-electric dipole antenna. In [11], a fully-integrated feature-rich 60 GHz SiGe BiCMOS antenna is developed and tested. In [12], a coplanar waveguide-fed broadband patch antenna is designed, microfabricated and characterized. A 15 % bandwidth and 5.5–7 dB gain are obtained. In [13], a new differentially-fed planar complementary antenna array is proposed relying on a low cost process. 25 % impedance bandwidth and 11.5 dBi average gain are achieved. In [14], a System-in-Package approach is used to address 60 GHz applications. A maximum gain value of 7.8 dBi is reached. In [15–17], a CMOS transceiver chipset is developed. Finally in [18], a 60 GHz monopole antenna with slot defected ground structure is presented.

As presented in this part, the WiGig technology is extensively explored in different researches, especially concerning the inherent beamforming techniques that have to be implemented to arise antenna gain at 60 GHz.

2.1.1.3 Li-Fi

Some recent works showed the possibility to make a connection by the Light Fidelity (Li-Fi) technology. Li-Fi is a "post-Wi-Fi" wireless technology based on the use of Visible Light Communication (VLC) (instead of radio frequency waves for Wi-Fi). Li-Fi is a 5G VLC system that uses light from Light-Emitting Diodes (LEDs) as a medium to deliver networked, mobile and high-speed communications. Li-Fi principle relies on the data transmission by amplitude modulation of light sources, according to a well-defined and standardized protocol. VLC works by switching bulbs on and off within nanoseconds which is too quickly to be noticed by the human eye. Li-Fi is different from laser, fiber optic or infrared communications by its protocol layers. The light waves cannot penetrate walls which makes a much shorter range (about a decade of meters, a few more than Bluetooth technology), though more secure from hacking, relative to Wi-Fi. A complete solution includes a standardization process, established by the IEEE 802 workgroup. Indeed, the Li-Fi relies on the IEEE 802.15 standard.

All big companies in electronics, such as Philips, Siemens or General Electric, work on this new communication technology [19, 20]. The French start-up Company, Oledcomm, is the most advanced on the subject. Two researchers of Versailles University, working on the technology since 2005, founded the Oledcomm Company in 2012. The start of the Li-Fi technology is a direct consequence of the migration of the light to the LED lamps, electronic devices that are suitable for high-frequency modulation. This technology allows throughput of several hundreds of Mbps and even up to several Gbps. As all new technology, Li-Fi has some drawbacks. The communication requires obviously a light on during the transmission. Furthermore, no mobility is possible.

SNCF takes an interest in the Li-Fi since several years [21]. Studies on geolocalization products in railway stations and transmission of information in trains via the reading lights were performed. The tennis competition of Roland Garros presented a connected TV via Li-Fi in June 2014. In a museum, when approaching a work, visitors can have access to an informative multimedia content. In a supermarket, the trolleys connected to the lighting system via a tablet can provide a range of services to clients. A large number of applications can be based on the Li-Fi technology, the different recent applications can be found on the site and blog on Li-Fi [19, 20]. However, one of the drawback is the fact that use cases are mostly unidirectional. For example in Museum, only downlink is possible and for supermarket the up-link is complicated, and is done via Wi-Fi. This one-way link is common for many use cases involving mass-market devices like computers.

Currently, a project between Luciom Company (people from Philips NXP) and the CEA-Leti is investigating a bidirectional Li-Fi modem allowing wireless Internet access up to 20 Mbps [19]. To our knowledge, there is no studies performed in the context of an Internet access on board trains via Li-Fi transmission in trains. However, the announced evolution by Oledcomm is to provide Internet access via Li-Fi. There is no doubt that it is a topic to investigate on board trains by performing transmission via individual lights of passengers.

2.1.2 *Classification of Technologies to Connect the Train to Internet*

Several technologies can be used to link the train to the Internet backbone. As mentioned in Chap. 1, a set of technologies are embedded in the TAT. Each technology can connect opportunistically. The criteria to select a particular technology among the available ones are typically the quality of the connection (signal strength), the delays, the throughput or the costs. The point of divergence between the different existing solutions that will be presented relies on the different technologies employed by the TAT and how it integrates them in order to provide a continuous connection.

We can consider two major families of technologies: satellite and terrestrial technologies. Satellite solutions can be based on different types of satellites (GEO, MEO, LEO), they can use different frequency bands and can be unidirectional or bidirectional. The following section gives all the details on the use of these technologies. The terrestrial solutions can be divided into two subcategories:

- Technologies relying on existing networks, so-called public cellular networks solutions;
- Technologies requiring the deployment of a specific ground-infrastructure: the dedicated train-to-infrastructure solutions. Among these are, in particular:
  - Leaky coaxial cables;
  - Solutions based on Wi-Fi or WiMAX;
  - Radio-over-Fiber;
  - All-optical solutions.

Several thesis and reports achieved a State-of-the-Art of existing systems to provide Internet on board trains [2, 22, 23]. Several papers detail specific systems developed to provide Internet access on board trains. The website of the railway operators were also used to find information. An annual conference dedicated to “Wi-Fi on Trains” takes place every year in London, the “TrainComms conference”. The papers presented at the conference helped to update the information about existing systems. The state of the art presented in this chapter relies on all of these studies and documents.

2.2 Satellite Solutions

2.2.1 *Description of the Technology*

Communication satellites represent a first solution to enable broadband Internet access on board trains. The main advantages of such solutions are [22]:

- The easy coverage of a large geographical area (one geostationary satellite can cover a quarter of the earth surface);

- The well adapted broadband connectivity for connection and aggregation of the traffic of a large number of mobile terminals;
- The resistance to high velocity;
- The low CAPital EXpenditure (CAPEX) due to the absence of installation of a dedicated infrastructure on track.

Nevertheless, the use of satellites leads to several constraints on systems design, which have to be taken into account [22]:

- Use of satellite requires satellite in Line-Of-Sight (LOS) in order to obtain broadband connectivity. Any obstacle between the satellite and the receiving antenna (catenary, bridge, high buildings) generates fadings or total loss of signal;
- Antennas require high antenna gain and a very thin beamwidth. It is then necessary to implement a precise tracking of the satellite. Moreover, train suffers of several movements, tracking solution of the satellite have to be even more precise in order to avoid interferences with other satellites;
- NLOS areas, such as tunnels, urban areas or stations, can lead to signal cut-off of several minutes and require the combination with other technologies, so-called "gap-filler". Two main solutions can be considered:
  - Satellite repeaters: one antenna is installed on the ground in order to recover the satellite signal and to redistribute it in the non-visibility areas. This kind of solution requires the deployment of an infrastructure along the track. Furthermore, specific authorizations have to be asked to railway infrastructure owner and to telecommunications regulator;
  - Vertical handover: allowing switching to other technologies, such as Wi-Fi, WiMAX or cellular networks (3G/4G).
- Railway constraints have to be obviously taken into account: Electromagnetic Compatibility with existing systems, installation, maintenance and space to install the antennas. Furthermore, satellite solutions can provide high throughput by using large antennas. Railway constraints force the railway operators to limit the size of the antennas, limiting then the delivered throughput, especially in the case of double deck trains.

Different kinds of satellites exist (cf. Appendix B.4): Geostationary Earth Orbit (GEO) satellites, Medium Earth Orbit (MEO) satellites and Low Earth Orbit (LEO) satellites. GEO satellites are generally very attractive because they use a geosynchronous orbit located at 36,000 km from the surface of the Earth, at equator level, which allows them to be seen as a fixed point in the sky. Moreover, GEO satellites cover a large geographical area and they are the only ones capable of providing broadband connectivity for mobile users. Thus, they are largely used in several existing communication and broadcasting systems. Satellites may have still some drawbacks. The use of GEO satellites leads to important propagation delays (around 400 ms) compared to MEO or LEO ones. This propagation delay may become a problem in the case of highly interactive applications. Modifications and optimizations are then necessary to accelerate the TCP/IP flow. Furthermore, GEO satellites being at equator level, north latitudes are then at weak elevation angles. This conducts to a reduced availability of satellite in case of obstacles. Finally, bandwidth has high costs (more than 1.5 M euros in Europe for a 36 MHz transponder per year).

Despite all these inconveniences, all connectivity solutions on board trains using satellite technology rely on GEO satellites. That can be explained by the fact that GEO satellites guarantee a large choice of products, constructors and satellite operators, together with a high capacity. Indeed, MEO and LEO are not able to provide broadband connectivity.

Another problem of satellite systems is a high OPerational EXpenditure (OPEX) due to the satellite capacity. Available throughput depends on satellite capacity; generated costs have to be taken into account in the business model. Nonetheless, clients desire more and more throughput, which arises bandwidth costs. An increase of the number of clients can generate an increase of incomes, but not an increase of throughput. Business model causes some big problems.

Satellites in Ka band can represent a solution to this problem because of their high capacity, which induces a reduction of bandwidth costs (3 to 5 less expensive than the Ku band). Moreover, satellites in Ka band operate at higher frequencies, which allows reducing the size of the antenna. The use of these satellites causes some problems yet. First of all, equipments in Ku band are not compatible with Ka band, which requires the development of new equipment fitting railway constraints. Moreover, signals in Ka band suffer of high attenuation in the case of bad atmospheric conditions (fog, snow, rain). These attenuations can reach 15 dB in worst cases. Finally, existing satellites in Ka band have a coverage area of about 250–500 km in order to allow a geographical reuse of frequency bands (and then optimize satellite capacity). A dynamic frequency allocation and a horizontal handover have to be implemented to assure connectivity of train from a cell to another. Global system will then be more complex. A complete study on Ka band still have to be performed, such as investigation on mobility effects and cell changes. These issues will be seen in the Chap. 3.

It is also important to notice that there have been recent developments regarding billing of bandwidth. Only bandwidth actually used is now charged, a "billing per us". Furthermore, the future is to use flat antennas that can be much more easily installed on trains.

2.2.2 Existing Studies, Projects and Solutions

2.2.2.1 Studies and Projects

Several studies and projects have been performed from satellite technologies. A survey on mobile satellite systems is presented in [24]. The report details the existing standards (such as S-UMTS, DVB-S2, DVB-SH) and the existing mobile satellite systems (such as Inmarsat, Globalstar, Thuraya). The different systems are then compared based on a number of criteria (frequency bands, PHY layer characteristics, multiple access techniques, satellite characteristics). A tutorial on satellite systems for Internet access is presented in [25]. More details of these two reports are given in Appendix B.4.

As presented above, when the satellite link is blocked, the TAT switch to a terrestrial network. This solution of "gap-filler" was first introduced in the ROSIN (Railway Open System Interconnection Network) project in 1999 [26], which intended to develop a system allowing supervising equipments on board trains thanks to a GSM connection between the train and the control center. The ROSIN project aimed to validate a complete and open platform, which represented the basis for a new generation of vehicles, consisting of an on board network that interconnects all various on board systems and subsystems. Works were pursued with the TRAINCOM [27] and the FIFTH [28] projects.

TRAINCOM [29] is a European project finished in 2003. 13 partners worked on the project, such as Siemens, Bombardier, Alstom, DB, and Trenitalia. During project life, two important railway operators, SNCF (France) and SBB (Switzerland), joined the TRAINCOM project as Observer Participants. The project aimed to develop a reliable communication system between the train and the ground, offering access to on board equipments and integration of all new available technologies (GSM or GSM-R links, protocols and language of Internet such as TCP/IP). The train was then connected to the ground with several wireless connections, and it could switch between them according to required bandwidth for a given application.

FIFTH (Fast Internet for Fast Train Hosts) project proposed a new network solution able to provide a broadband Internet access to passengers on board HST via satellite solutions. A new satellite technology was studied and a prototype was designed and developed in order to implement a practical demonstrator. The prototype was based on two subsystems: the railway mobile terminal and the network access infrastructure. The railway terminal was composed of the satellite network access interface and all the subnetworks in the train for passengers (servers and users terminals). Tracking and pointing techniques were based on a GPS navigation system and an inertial technique (gyroscope). A bidirectional satellite solution used in "classical trains" (non high speed) provided a communication with throughputs of about 2 Mbps/512 kbps (download and upload respectively).

The solution was then integrated under the INTEGRAIL project [30, 31], in the context of an intelligent integration of railway information system. Other works evaluate the TCP flows of satellite systems [32]. Finally, the TRAINIPSAT project [33] aimed to define, specify and test a technical solution to provide connectivity services for HST, both for individuals and professionals. The objective of the project was to demonstrate the feasibility and relevance of a solution combining a bidirectional satellite link and a terrestrial link, and a seamless connectivity on board train via a network such as Wi-Fi. The terrestrial link, based on the WiMAX technology, was specifically designed to take into account technical constraints due to fast mobility. The satellite link was explored with the development of a predictive model of availability of satellites, relying on Markov models. Mobility management and handover mechanisms were also investigated.

Some experiments were performed in Spain in the AVE trains (High Speed Trains) of the RENFE with Indra. Indra [34] is a multinational located in Spain and Latin America. It provides solutions and services in different domains, such as transportation, traffic, energy or industry. Indra experimented a solution to provide broadband Internet access on board trains. The system is based on a bidirectional satellite connection using Demand Assigned Multiple Access (DAMA) access scheme in order to optimize the use of the frequency band. Frequency bands are then automatically assigned to mobile terminals, depending on their needs. Indra's system manages three satellite technologies: DVB-S for downlink, wide spectrum Code Division Multiple Access (CDMA) for uplink, and Single Channel Per Carrier (SCPC) for both links, coexisting with SCPC and/or CDMA. Test measurements were performed on the line between Barcelona and Madrid. No further information could be found on these trials.

2.2.2.2 Developed Solutions

Two main companies provide solutions based on satellite technologies: Icomera and 21Net.

Icomera

Established in 1999, Icomera is headquartered in Sweden with office in the United Kingdom and channel partners worldwide. Icomera's products are deployed on rail, road and sea. Icomera developed a multi-technology platform using satellite technology for the downlink and cellular technology for the uplink, in order to provide broadband Internet access in trains with Wi-Fi deployed in the carriages. For the railway context, Icomera's solution relies on the X6 platform. In 2014, Icomera system was enhanced to be able to access LTE technology. The system comes with four LTE modems and Wi-Fi capability plus an additional modem or Wi-Fi slot for future expansion. Each modem slot has two SIM card slots and supports geo-fencing SIM card selection allowing operators to reduce costs in cross-border scenarios [35]. Throughputs can then reach 40 Mbps [36].

First tests of broadband on board trains in the world were performed in Sweden in September 2002, with the first deployment in January 2003 with Scandinavian rail operator Linx (owned by the Swedish Company SJ and the Norwegian Company NSB), between Gothenburg and Copenhagen, using Icomera platform. Since 2005, SJ offers Internet on board the whole network of Intercity and commuter trains [37].

The Icomera platform is also used since 2004 by Intercity East Coast Railway franchise in UK running from London to Scotland (operated by GNER, then National Express East Coast and currently East Coast Railway companies). East Coast trains carry up to 500 passengers at speeds up to 200 km/h through 400 miles of urban, suburban and rural areas. A single antenna is installed on the roof of the train and the different carriages are linked using the train lighting circuit. The main used technology is the satellite; system switches on cellular technologies in case of non-visibility [2]. Initially, the system was based on the combination of a satellite link and a GSM link, allowing average throughputs of 0.5 Mbps. Then to arise performance, several cellular 3G/High Speed Packet Access (HSPA) networks can be used at the same time (until 8). East Coast fleet is currently being upgraded with the new Icomera system.

From 2010, Icomera and Fleetconnect associated to install passengers Wi-Fi on board Irish Rail trains. Fleetconnect is an Irish provider of public transport Wi-Fi services. The system uses multiple 3G+ mobile broadband networks to deliver a fast availability connection. Icomera was awarded the contract to install Chiltern Railways mainline fleet in UK in early 2011. Fleet installation was completed for Chiltern mainline service, launched in September 2011. The Icomera mobile application router is at the heart of the Chiltern Railways Wi-Fi service, and uses multiple HSPA mobile broadband networks to deliver a fast availability. The system is ready to take advantage of new faster 4G services as these are rolled out in the UK [35].

In Czech Republic, the Czech Railway company contacted Icomera in the late 2011 in order to improve the level of services offered to its passengers on board its Pendolino trains between Prague and Ostrava. The aim was to provide connection to passengers via Wi-Fi and also to provide additional entertainment options via infotainment system. The solution is then based on the Icomera platform in partnership with Simac passenger infotainment system [35].

ScotRail, the national Scottish railway company, awarded Icomera to the contract to provide on-train Wi-Fi services on its trains running from Glasgow to Edinburgh. Initially announced on December 2012, the installation was completed in late 2013 [35].

Finally, a contract was won in October 2014 by Icomera to supply on-board Wi-Fi to the fleet of vehicles operated by Dutch transport Rotterdamse Elektrische Tram (RET). The installation of the complete information system and Internet on board was expected for December 2014 on the fleet composed of 113 trams and 145 metro trains.

It has to be noticed that Icomera got the contract to renew the Thalys connectivity, from satellite to cellular. Icomera is no more focused on satellite but also on cellular solutions.

21Net

21Net is a British Company founded in 2002. It received support and funding from the European Space Agency (ESA) and the British National Space Centre (BNSC). In 2004 in the context of ARTES project (2004–2006), 21Net set up trials with Spain's national rail operator RENFE demonstrating access to broadband Internet via a bidirectional antenna on a HST running at over 300 km/h, allowing throughputs up to 4 Mbps/2 Mbps (downlink/uplink) [38]. The satellite solution relies on the Hispasat satellite, DVB-S technology for the forward link and SCPC for the return link. The "gap-filler" relies on cellular solutions. An upgraded cellular solution is implemented since mid-2013 relying on a multi-operator and multi-SIM bonding that aggregates bandwidth across multiple channels simultaneously. The system is based on MIMO techniques, recent cellular technologies (LTE, HSPA) and standard 2 bonded SIMs per operator, ideally using all available networks [39]. The satellite solution was also upgraded in order to increase spectral efficiency and availability with Adaptive Code Modulation (ACM) [39]. An optimal usage of available bandwidth is also implemented, such as an advanced accelerator and proxy optimized for mobile environment, a fair bandwidth distribution among passengers and a blocking of services [40]. 21Net is currently working on new satellite flat antennas in Ka band [38]. The system is deployed in different railway contexts.

In 2005, 21Net, in collaboration with Nokia, runs a commercial pilot train on the Thalys network to deliver broadband Internet access in its HST, combining a satellite and a cellular links [41]. The overall fleet of 26 trains was equipped with the ThalysNet system in October 2008. With its mobile access router, the 21Net system combines and aggregates several cellular links and a satellite link thanks to a satellite antenna in Ku band set up on the train roof. A single DVB-RCS modem was developed to share bandwidth among all the trains in the network, and to allocate band on demand depending on needs. Throughputs up to 4 Mbps/0.5 Mbps are recorded [38]. In November 2014, Thalys launched a trend to update its Internet on board service. Specifications are to rise throughputs 5–8 times [42]. Icomera won the call for tenders and will equip the Thalys fleet for the end of 2015.

In 2009, NTV (Nuovo Trasporto Viaggiatori), an Italian railway Company chose 21Net to operate the entire Telematics system (Broadband Internet Multimedia Entertainment) in their HST. For this project, 21Net worked with Alstom in order to integrate the system in the design of the 25 AGV (Automotrice à Grande Vitesse). Satellite antennas were then perfectly integrated in the AGV trains. First equipped trains started in May 2012 and the entire fleet was equipped in February 2013. The system relies on the combination of a bidirectional satellite link and several cellular networks. The system is also equipped of a multimedia portal with touchscreens, live TV, VoD, newspapers, books, etc. Average throughputs than can be reached are 8 Mbps/0.5 Mbps [38]. Currently, NTV is migrating from a satellite solution to a cellular one. 21Net is responsible of the rollout, monitoring and integration of the new technology [40].

In January 2009, 21Net and Techno Sat Comm performed tests on the lines of the Indian Railway operator [43]. Three Radjani Express trains are then equipped in February 2013 on the line between Delhi and Calcutta. The system operates also from a bi-directional satellite link and several cellular networks. Seamless broadband connectivity of 9 Mbps at 180 km/h were recorded. The satellite solution is scalable to 3rd Generation Ka band systems that will allow throughputs up to 1.5 Gbps for the downlink.

Other Solutions

PointShot Wireless is a Canadian Company built in 2002. It provides a number of wireless solutions for broadband connectivity. The RailPoint solution was developed for the special case of broadband Internet access on board trains. This solution was deployed from 2006 in the Via Rail trains in Quebec. Connection between train and ground was established using satellite, cellular networks (GSM, GPRS or UMTS) or terrestrial links (Wi-Fi, WiMAX). To our knowledge, no further information are given about the precise technical solutions used and it seems that the PointShot Wireless Company went out of business. Current solutions in Via Rail trains are presented in the Sect. 2.3.

Temir Zholy, the national railway company of Kazakhstan, equipped its Tulpar HST with an Internet on board access in 2011. Gilat's Very Small Aperture Terminal (VSAT) platform was installed on the trains. Gilat Satellite Networks is a public company headquartered in Israel that develops and sells VSAT satellite ground stations and related equipment [44, 45]. The Gilat system relies on GEO satellites. Throughputs up to 2 Mbps can be obtained.

Zoom on SNCF Solution

The French railway Company SNCF performed several tests [22, 46, 47], in order to provide a broadband Internet access on board HST. Experiments on train-to-infrastructure solutions were performed and will be presented in the Sect. 2.3. Combined solutions with satellites for the downlink and cellular networks for the uplink were also tested. Radio cellular coverage being too weak in France (contrary to some countries as Sweden), works initially focused on bidirectional satellite solutions [47]. Two solutions were tested: the Thales Alenia Space and the 21Net ones. These two solutions are based on DVB-S technology for downlink and SCPC for uplink. However, Internet applications on board trains require flexible frequency allocation, together with multiple access techniques able to distribute the different operated trains at a given moment. SCPC solution remains too inflexible at this level. Researches interested at the DVB-RCS technology. A first Internet service on board trains was launched by SNCF in December 2010 on the TGV-East line, the "BoxTGV". The system relies on a bidirectional satellite solution on Ku band (frequencies of 11 and 14 GHz, for downlink and uplink respectively) with a Wi-Fi coverage and 3G for NLOS areas (the "gap-filler" solution). Orange Labs Company and Alstom (supplier of the on board hardware bearing Orange software) were involved in the research and development of this solution [48]. However, the system never found its profitability (expensive technical architecture in terms of CAPEX and OPEX). In addition, technology could not be deployed on "Euroduplex" with two levels for technical reasons (high railway constraints), which leads to a poor legibility of the offer (offer not available on all trains). The "BoxTGV" system was then stopped in December 2013. Other studies are currently on going at SNCF, as seen in Chap. 3.

2.2.3 Summary on Satellite Solutions

As presented in this section, several "Internet on board trains" solutions relying on satellite technologies are deployed in the world. It has to be noticed that all these solutions rely also on cellular solutions, acting as "gap-filler". Nevertheless, this kind of solutions remain expensive and provide limited throughputs, as illustrated in Table 2.2. Currently, the trend is globally to go towards less expensive solutions, such as public cellular networks solutions, as presented in the next section.

However, a solution to enhance the systems is to use Ka band, as presented at the beginning of this chapter. Indeed, Ka band provides higher satellite capacity,

Table 2.2 Summary of throughputs of existing satellite solutions

Deployed solution	Downlink throughput	Uplink throughput
East Coast Railway	40 Mbps (announced)	/
Thalys	4 Mbps	0.5 Mbps
NTV	8 Mbps	0.5 Mbps
Indian Railways	9 Mbps	/
Temir Zholy	2 Mbps	/

which reduces costs and increases performance. Nevertheless, studies on Ka band still have to be performed, such as investigation on mobility effects and cell changes. This solution is envisaged by Indian Railway, which announced throughputs up to 1.5 Gbps.

In the long term, other types of satellites could be investigated, such as nanosatellite, which belongs to miniaturized satellites. Some future satellite technologies are presented in Chap. 3.

2.3 Terrestrial Solutions

2.3.1 Public Cellular Networks Solutions

2.3.1.1 Studies and Project

The public cellular networks solutions are usually based on the use of several public cellular networks deployed over landmasses. The TAT integrates several links (up to 8 in some cases) with different Mobile Network Operators (MNOs). Thereby, the TAT can manage the lack of coverage of one operator by supplying it with another one with better coverage. In the case of no coverage at all, a "gap-filler" solution can be used, as for the case of satellite solutions. Several solutions were deployed using cellular solutions. In [49], some field measurements are performed with the MNO 02 in UK. The objective was to evaluate TCP performance. The results show an average throughput at TCP level of 30 kbps with GPRS and 340 kbps with HSDPA. The authors pointed out the huge contrast with the theoretical throughputs announced by the operator (56 Mbps for the downlink, 22 Mbps for the uplink). In [50], two mobile Internet Service Provider (ISP) are compared in Korea. Each operator is operating two networks: 3G and 3.5G. Two scenarios are considered: an HST at 300 km/h and a mobile car at 100 km/h. Throughputs around 500 kbps at UDP level and 1 Mbps at TCP level, both in downlink and for both 3G and 3.5G are observed.

Nomad Digital [51], a specialist in on-vehicle ICT, provides wireless solutions to the transportation sector: trains, metros, trams and buses. The Company developed a wide range of solutions based on a scalable on-board IP platform, allowing passenger Internet access on board via Wi-Fi, and local contents, such as passenger information, infotainment and displays. The main solution is the NDConnect mobile router used in many rolled out solutions in the railway domain. Nomad Digital provides solutions based on public cellular networks only, and solutions based on the combination of cellular and WiMAX technologies. They are presented in the following.

2.3.1.2 Solutions Based on Public Cellular Networks only

The East Midland trains in UK are equipped with an on-train Wi-Fi since 2011, relying on the Nomad Digital system [52]. The system was upgraded in 2014 in order to improve connectivity speed and reliability.

DSB, the state-owned Danish rail operator, decided to equip all its Metropolitan S-trains in Denmark's capital Copenhagen with wireless communications, after a study revealed that real-time traffic information was the number one request from its daily passengers. The survey revealed that even in the event of delay, complaints would be minimized and customer satisfaction raised by providing accurate, up-to-the-minute information on new times of arrival, connecting traffic and service alterations. Free Internet access using the same wireless communication system was built into the package to further increase customer satisfaction. The communications between train and ground are provided by an NDConnect Communications Control Unit (CCU) mobile router from Nomad Digital, aggregating two mobile networks. The solution was chosen to be scalable with a modular approach in order to support new technologies and standards, such as LTE. Access points in the carriages and dedicated portal and infotainment servers support the services.

NSB, the Norway's national rail company, has implemented wireless Internet access for passengers on its intercity train's fleet. The country is large and sparsely populated outside major cities, which implies that mobile broadband coverage is patchy and frequently blocked by tunnels. Nomad's multi-carrier aggregating NDConnect solution presented above is used. The system exploits all public networks in Norway and a particular requirement was to use the ICE CDMA network operating at 450 MHz. The solution is scalable and modular to fit with LTE standard, which was launched extensively in Norway.

The NDConnect system is also used to provide passenger information and Internet access in the intercity trains of the NS Dutch Railways, in Netherlands. NDConnect router uses national cellular networks. Recently, NS dutch Railways announced that the entire Dutch intercity fleet will be equipped with 4G mobile internet connections by the end of the summer of 2015. A fair usage policy will be also initiated to boost Wi-Fi speed for all passengers, by limiting the speed per user to 150 kbps.

Queensland Rail, one of Australia's largest train and transport companies, which operates around 200 commuter and regional services along 7000 km of track, is rolling out free Wi-Fi on its trains. Some parts of the country suffer of poor coverage, which means loss of communication. The developed system works then across multiple networks. Nomad's technology uses the three main MNO carriers, which guarantees a higher level of network coverage and better bandwidth availability.

In [53], the author announces that Eurostar aims to bring Wi-Fi on board its fleet. Nomad Digital won the contract to provide the on board technology. Indeed, Nomad Digital is able to aggregate bandwidth from different MNOs, as explained in this part. A crucial point is also that it enables cross-border connections, which is important given that Eurostar currently across three countries (France, Belgium and UK) and plans to extend its reach to Germany and to the Netherlands.

Other solutions, which do not use Nomad Digital system and based on public cellular solutions, were also rolled out in different countries.

In Canada, the Via Rail trains, formerly equipped with the PointShot Wireless system, are now providing Internet access on board from three different wireless providers [54]. Eight antennas are mounted on the front-most cars of each train set. Trains from Quebec to Windsor and Montreal to Halifax are equipped with this solution.

In Denmark, a cellular solution was deployed in the Arriva's train, relying on an Icomera solution [37].

In Latvia, Latvian Railway, in cooperation with the wireless telecommunications company Triatel, provides an Internet services on board its trains since 2009 [37].

In Switzerland, the Swiss Federal Railways, the national railway company, proposes modern signal amplifiers in their trains to ensure better reception for passengers on the train. The Company announced that all long-distance trains will be fitted with 3G/4G signal amplifiers by the end of 2014. The signals came directly from the mobile phone signal from outside and are amplified into the coach.

Initially, the RailNet service on board ICE trains in Germany integrated 3G networks with a Flarion FLASH-OFDM based network. The T-Mobile mobile phone operator deployed its network as a "gap-filler", but soon after it was demonstrated as a feasible solution, so the coverage was extended [55]. Currently, the Telekom Company is responsible for the on board system and the trackside network, and also for the ISP. Telekom gets the exclusiveness on Wi-Fi in ICE trains. DB Company buys then "online-minute". Telekom manages operational, mobile networks, server on train and connection between coaches.

In Hungary, the Gysev Railway Company equipped its trains from Budapest to Sopron with a free Wi-Fi Internet access on March 2011 [56]. The system relies on the Telenor Telecommunication Company system. It uses High Speed Uplink Packet Access (HSUPA)/HSDPA/(Wideband Code Division Multiple Access) W-CDMA networks. Announced throughputs are 7.2 Mbps for the downlink and 5.76 Mbps for the uplink. Recently, the Russian Railways' subsidiary Aeroexpress announced that free Wi-Fi services will be available on all its trains running between the city center of Moscow and the airports. The broadband wireless link is provided by RTD-Telecom using 3G and 4G networks belonging to the main Russian mobile operators: MegaFon, Beeline, MTS and Yota. 20 to 25 Mbps average throughputs are announced. Moscow's metro trains were previously equipped with Wi-Fi connection on board, with 90 Mbps announced throughputs.

2.3.1.3 Combination of Public Cellular Networks and Dedicated Infrastructure

Southern Railway is a train operating company along 80 km of track in the south of England. Nomad Digital developed an on train broadband switching solution incorporating both 3G and 802.16 (pre-WiMAX) technologies to cover the entire Brighton Mainline route. Nomad Digital worked with T-Mobile in order to develop the system. Throughputs up to 2 Mbps can be obtained. The service seems to have been stopped since 2011.

Heathrow Express is the service to and from Heathrow Airport in London. Nomad Digital took T-Mobile as a partner in order to offer passengers Wi-Fi connections on board trains up to 2 Mbps. T-Mobile built an optimized wireless network along the entire line. On the 15 min of the trip, 6 km are in tunnel. This area is then covered by WiMAX radios. Five WiMAX ground stations covers the entire tunnel. The train in then equipped with three antennas: two for WiMAX and one for HSPA. Passenger accesses are provided through IEEE 802.11g; links between cabins rely on IEEE 802.11a. Since 2013, the system is being upgraded with an updated WiMAX in tunnel and a Vodafone 4G connection bonding 6 SIMS. Currently, the link between on board Wi-Fi and available Wi-Fi in the different terminals of the airport is performed [57]. Future uses are planned by Heathrow Express, such as staff smartphone and tablet applications or enhanced ticketing [57].

As for the Southern Railway and the Heathrow Express, the Pendolino fleet of the Virgin Trains in UK were equipped with a system combining 3G and WiMAX technologies in order to provide on board broadband connectivity. Nomad Digital resigned a contract with Virgin in 2014 in order to boost the existing solution. 76 trains will be upgraded—56 Pendolinos and 20 Super Voyagers. The Pendolino trains will then be able to deliver service speeds of up to 12 Mbps, while the Super Voyager trains will be capped at 8 Mbps. Nomad's wireless routers will connect mainly to the 3G network, and they will be ready to take advantage of 4G connections when available.

Nomad Digital proposed its mobile router, able to switch between cellular solutions such as 3G and WiMAX technology deployed along the track to the UTA trains of Utah, in US. Recent news showed that the system encounters many problems such as harsh environment and very large number of potential users, leading to connection problems and low throughputs.

A partnership agreement between the telecommunication company Du and the RTA (Roads and Transport Authority) was established to provide Wireless Internet access on board the Dubai metro in the United Arab Emirates (UAE) from 3G networks and WiMAX technology, relying on Nomad Digital system. The NDConnect router aggregates both Du's network, HSPA and WiMAX (802.16e) at 2.5 GHz. Du is also responsible for the rolling out of an Internet access on board the soon-to-be-launched Dubai Tram service.

Today, Nomad Digital does not propose anymore solution based on WiMAX, but reversely is able to propose a solution based on a mix of cellular and satellite technologies (with partners).

Other solutions based on the combination of cellular and WiMAX solutions, but not based on Nomad Digital system, are also deployed, such as the Taiwan High Speed Rail (THSR), which equipped its trains with a system relying on WiMAX and 4G networks to provide connectivity on board since 2012. First studies are presented in [58]. The method uses a Distributed Antenna System (DAS) with Radio-Over-Fiber. Base stations can then cover wider areas with less interference. It is showed that the coverage can be extended from 3/4 to 17 km.

In US, Amtrak, the national Railway Company, equipped 85 % of its trains of an Internet access. Amtrak is composed of Intercity trains, such as the California fleet, the Amfleet Northeast Corridor, the Acela Express and the new Midwest service (since February 2014) [59]. The solution, based on cellular networks, was upgraded to 3.5G in 2011/2012 and to 4G in 2013, allowing throughputs up to 10 Mbps. On June 2014, Amtrak Company [60] announced an improvement of the existing service allowing throughputs up to 25 Mbps. To support very high speed application such as video streaming, VoIP, video conferencing, an average throughput superior to 3 Mbps per passenger is required. In order to achieve this goal, Amtrak is currently thinking of the roll out of a dedicated trackside network, based on the Fluidmesh system [59], detailed later in the section.

2.3.1.4 Summary on Cellular Solutions

As presented in this section, many rolled out solutions in the world rely on cellular solutions. Nomad Digital represents the most deployed solutions. It is also important to notice that Internet on board train is a very fast evolving subject. We perform a survey on cellular solutions, that can not be exhaustive because of new solutions appearing constantly. This kind of solutions can be deployed "alone" or combined with other solutions, such as WiMAX technologies. It can be noticed that it is quite difficult to obtain precise informations on the performance of the systems, in terms of throughputs especially. As presented in Annexe C, using actual public 4G, throughputs cannot exceed 30 Mbps. In [61], authors claim that throughputs cannot exceed 10 Mbps for Internet on board train solutions. However, at the TrainComms conference standing in London on June 2014, Icomera claimed that throughputs up to 250 Mbps can be reached relying on cellular-based solutions, depending on LTE deployment in the countries.

Cellular-based solutions are many deployed for Internet on board access because they allow low costs, relying on the use of existing infrastructures. However, cellular-based solutions lead to many drawbacks. Minimum capacity requires multiple cells management. Moreover, base stations are not often near the tracks, and antennas are not oriented for track coverage. Cellular-based solutions have then the main drawback of no control over Quality of Service, by depending on MNOs.

Joint works between railway and MNOs skateholders are currently used to implement strategy for better on train Wi-Fi services and better railtrack coverage. This is the case in Denmark and in France for instance.

2.3.2 *Dedicated Train-to-Infrastructure Solutions*

This kind of solutions consists of the rolling out of a dedicated infrastructure on the ground, allowing connectivity to the TAT. This connectivity can be obtained: with guided waves through leaky cables, with radio waves in free space via systems relying on Wi-Fi or WiMAX technologies, or Radio-over-Fiber technology, or with optical signals, such as full-optical systems based on lasers or diodes. All these solutions are described in this part.

The dedicated train-to-infrastructure solutions allow meeting growing demand in terms of throughputs. However, this kind of solutions rely on long, complex and expensive deployments of infrastructures along the track. In order to reduce costs, it is then essential to try to minimize the number of sites required to ensure radio coverage of the network. The radio coverage is thus one of the main features to be taken into account for the choice of the communication technology [62]. Furthermore, another key feature is the throughputs that the system allows to reach. These two features are closely related and the best tradeoff between range and throughput have to be found to optimize the system. Other features have to be considered also, such as reliability, security and need for licensed radio spectrum.

2.3.2.1 Radio-Based Solutions

In the early 2000s, Gavrilovich [63] and Lin [64] studied the problem of providing broadband communications to fast moving users. Gavrilovich [63] argued that a large number of small cells operating at high frequencies was the most economical and practical infrastructure for providing wireless broadband access to a large number of users. The model relies on moving base stations that travel along a track. These ones are then linked to fixed base stations via wireless links. The fixed base stations are uniformly deployed on the track. The combination of moving and fixed base stations allows broadband wireless communications with fewer handoffs. However, the moving base station concept may not be practical. In [64], another architecture is proposed for providing communications and entertainment on board high speed transport systems. An architectural design is discussed at the conceptual/functional level of communication and entertainment services on board high speed transport, such as HST, cruise ship or airplane.

Leaky Coaxial Cables (LCX)

In Japan, some authors [65] demonstrated a broadband Internet access on board trains from leaky coaxial cables (LCX). The system requires a cut-off management between the different segments of the leaky cable at high speed. Authors proposed a communication architecture for bullet trains (Shinkansen trains from Tokyo to Osaka), which consists of a base station with an Ethernet interface, and mobile devices. First test beds were performed and showed a throughput up to 768 kbps.

The Wi-Fi access on board "bullet train", running on the Tokaido-Shinkansen line, is now available since March 2009, based on the LCX technology [66]. The

NTT Communications Company provides the service. Theoretical throughputs up to 2 Mbps can be reached for the downlink, and 1 Mbps for the uplink. The error rate is less than 10^{-5} with error correcting codes [67].

However, such solutions require the deployment of the cable along the track, which leads obviously to high CAPEX and OPEX. Furthermore, a LCX system is non scalable. Once a frequency is chosen, no modification can be brought.

WiMAX

IEEE 802.16 is a series of wireless broadband standards, known under the name "Worldwide Interoperability for Microwave Access" or WiMAX. The main standards are summarized in Table 2.3. WiMAX's bandwidths and range capabilities make it suitable for a variety of applications such as providing portable and mobile broadband connectivity. Its range capabilities also make it an alternative for cellular phone technologies. Furthermore, its bandwidth capacity makes it suitable for not only providing broadband Internet access but also providing additional services such as VoIP. Technical details on the Mobile WiMAX IEEE 802.16e can be found in [68].

All the solutions based on the combination of cellular technologies and WiMAX developed by Nomad Digital were already mentioned earlier in the part dedicated to the cellular solutions: the Southern Railway of Brighton, the Heathrow Express, the Virgin Trains in UK, and the UTA trains of Utah in US. In these solutions, the WiMAX base stations were connected to T-Mobile network through ADSL uplinks at 2 Mbps, which in fact represents a bottleneck for the users since the WiMAX technology can reach 48 Mbps. The implementation of WiMAX technologies in the railway context seems not to exploit all the capacities of the WiMAX technology [23]. Nevertheless, the major advantage is the large coverage, of about 5 km.

In the literature, few papers present experimental analysis using WiMAX technology in the railway domain. Aguado [69] present an architecture providing broadband wireless communication on trains, able to address the security, performance and communication needs. Enhancements in mobility management were introduced in the WiMAX network. In [70], the authors propose a mathematical model to estimate the bit error probability of a WiMAX system in order to offer the best chance to achieve improved throughput with the high mobility. In [71], a state of the art on handover mechanism for WiMAX is presented. It pointed out that while WiMAX is a promising technology (in terms of QoS, bandwidth and costs), there are still open issues about the handover management.

Table 2.3 Main standards of the IEEE 802.16

Standard	Standardization date	Frequency (GHz)	Theoretical throughput (km)	Range
IEEE 802.16d	2004	2–11	75 Mbps	7
IEEE 802.16e	2005	2–6	30 Mbps	3.5
IEEE 802.16m	2009	–	1 Gbps (fix)/100 Mbps (mobile)	–

A WiMAX solution is deployed in the Narita Express train connecting the Narita airport to the city center of Tokyo in Japan, which represents 90 km running in 55 min [66]. The service started in October 2009. It uses a WiMAX technology at 2.5 GHz bands. Maximum throughputs of 40 Mbps can be obtained for the downlink [72]. Since 2012, the same system equipped the Super Hitachi trains, which are limited express trains running from Tokyo to Iwaki (200 km, 2 h). Solutions based on WiMAX are also currently studied for the Caltrain of Silicon Valley in US, relying on a Nomad Digital solution. No further technical details were found on this system.

Wi-Fi

Wi-Fi was presented in the section dedicated to the way to connect users to the TAT. In this section, we present the opportunity to use the Wi-Fi standard to connect the TAT to the global Internet. Wi-Fi network has then to be deployed along the track. Wi-Fi technology is a very interesting candidate among terrestrial technologies. Indeed, it is an unlicensed and well known technology allowing good performance and resistance to the high velocity.

Some works present results on evaluation and testing of the applicability of Wi-Fi to provide connectivity to trains. These works were performed with the Federal Railroad Administration (FRA) office in US [73, 74]. In [73], the tests performed showed that the 802.11b technology is able to establish a communication with a train up to 144 km/h. A throughput of 6 Mbps is obtained, but with variations observed due to handover issues. The average delay observed from the train is 40 ms. In [74], a model is implemented to evaluate the performance of 802.11b in an underground scenario. In these two works, the main highlighted drawback is the difficult management of the handover mechanism, which decreases the overall throughput of the system.

In [75], Bit Error Rate (BER) analysis are presented, confirming the same conclusions, especially on handover issues. In [76], similar measurements were performed but on an architecture providing Internet access to mobile users in vehicle along the road. In [77], the authors present results on measurements of Wi-Fi connections between an in-motion vehicle and an access point located on the side of the road. In all these works, the common issue is the handover mechanism. Wi-Fi technology is not well suitable for train mobility scenario, using IEEE 802.11b standard. Furthermore, the deployment of such an architecture would induce high costs for large network of access points along a railway. Finally, in [78], experimental results on throughput, delay and coverage range of both the Wi-Fi (802.11b/g at 0.9 and 2.4 GHz) and the WiMAX (1.5 and 3.5 GHz) technologies in a tunnel. The measurements showed some good results for the Wi-Fi IEEE 802.11b/g at 0.9 GHz, closed to the results for WiMAX at 1.5 GHz. Throughputs up to 22 Mbps are obtained. However, WiMAX technology suffers of higher delays (around 35 ms) compared to the Wi-Fi one (around 25 ms).

SNCF, the French National Railway Company, performed, in collaboration with Orange Labs, some experimental tests relying on Wi-Fi IEEE 802.11b and g [47]. The tested network was based on 4 access points located on bridges and pylons, covering an area of 13 km in Vendome, near Tours in France. Connectivity performance tests were performed showing a network able to support a 2 Mbps traffic along the handover across the 4 access points. An extended network of 50 km was then deployed relying on 10 access points installed on 3G cellular sites. Results show some good performance in terms of data transmission throughputs. More recent experiments were performed in 2010 using the IEEE 802.11n standard [22, 62]. Two base stations were placed at 6.3 km from each other, close to the average distance between two consecutive GSM-R sites in France. Throughputs up to some tens of Mbps were achieved.

To our knowledge, only one real system was deployed. Indeed, the Tsukuba express in Japan provides Internet connectivity in its trains since 2006, based on the Wi-Fi technology [66, 67, 79]. Throughputs up to 54 Mbps are indicated but no further information could be found.

It has to be noticed that Even if the Wi-Fi technology was not designed for handover initially, several proprietary solution have been built by Signaling suppliers on the top of 802.11 standard, in order to provide efficient mobility. Wi-Fi train-to-ground connectivity is widely spread for metro segment bearing both CBTC and broadband services.

For instance, recently, Madrid metro starts experiment on Wi-Fi based solution with large fleet (around 2300 cars, 300 km of tunnel and 280 stations) with 8000 Wi-Fi Access Point deployed (4800 of them on board), 5000 IP HD-cameras, 800 train-to-wayside base stations...

Radio-Over-Fiber

"Classical" cellular networks have the main drawback in the case of high mobility: frequent handovers during the passage from a base station to another, which leads to a significant decrease of throughputs. One solution to this problem is to deploy a system based on Radio-over-Fiber (RoF).

In [80], authors argue that broadband connectivity can be obtained by reducing the size of the cells. However, it leads to the roll-out of a large number of base stations along the track. The authors propose then to use a RoF system, allowing feed base stations deployed along the track. Antennas fed by optical fiber are called Remote Antenna Units (RAU). The goal of a RoF system is to transfer complicated signal processing functions from the base stations along the railway to a centralized control station, and then to reduce costs deployment and frequency of handovers. For communications between the access network and the train, data are modulated at control station level and sent into optical format to each RAU, using multiplexing on wavelength, each RAU having a unique wavelength for communications. The antenna transforms the optical signal into a radio signal transmitted to the train. For communications from the train to the access network, the closest antenna captures data. To reduce handoff times at train access terminal, [80] propose to use the concept of "mobile cells". It consists of a cell reconfiguring constantly at the same speed than the train, so that the train access terminal communicates on the same frequency the entire ride.

The Radio-over-Fiber technology is used in the Shanghai Transrapid, which is a MAGLEV train running up to 500 km/h. This train runs between the Shanghai airport and the city center (around 30 km long). The system was implemented by the Telefunken RACOMS German Company and uses a communication system relying on fiber optic links and radio base stations deployed along the track of the train

[23]. Throughputs up to 4 Mbps can be obtained in full duplex at 3.5 GHz and up to 16 Mbps in full duplex at 5.8 GHz [81].

Other Proprietary Solutions

In addition to technologies, such as Wi-Fi or WiMAX, other solutions can be used to provide Internet on board trains, the proprietary solutions such as Fluidmesh, Luceor and Reicom.

Fluidmesh [82] was founded in 2005 by a team of researchers from the Massachusetts Institute of Technology (MIT) and the Polictecnico of Milan. Their goal was to reliably deliver fiber-like performance via unlicensed wireless spectrum, providing connectivity for transmission of video, voice, and data. Fluidmesh aims to bring broadband connectivity to sites and environments that are today too hard or large to connect, such as high speed moving vehicles and trains, large-scale industrial sites, distributed infrastructures and complex urban environment.

Fluidmesh developed a transmission protocol called FLUIDITY™ which is a license-free trackside wireless, operating in the 5 GHz band. They claimed to provide broadband up to 100 Mbps on a train running up to 350 km/h without service disruption. Furthermore, handoffs below 3 ms are announced. The system relies on a 2×2 MIMO-based radio technology and dual-polarized trackside and on board antennas. In fact, it consists of a modified Wi-Fi.

In June 2014, Amtrak announced its collaboration with the Fluidmesh Company to install dedicated trackside technology on its HST line connecting Boston to Washington, D.C.

Luceor [83] is a French Company founded in 2005. Luceor is a specialist in outdoor wireless networks developed for 3 main applications: emergency situations (natural disasters, industrial accident), events (political meetings, cultural or sporting events) and infrastructure (industrial sites, public transportations). The solution is based on the WiMESH technology. WiMESH is a routing and mesh technology based on IEEE 802.11n standard. It allows wireless devices to connect to the next, in a dynamic and instant way with no central hierarchy to form a "mesh" structure. Throughputs at UDP level up to 450 Mbps are announced, and mobility is supported until 350 km/h. Furthermore, coverage from 100 m to 10 km are announced.

RATP performed some tests in June 2014, based on the WiMESH technology. Experiments were in the context of tunneling emergency situations, needing new generations of wireless transmission of video streams and voice in mobility and very high speed, such as Luceor equipment.

Reicom [84] is an Italian Company established in 1999, that develops products for broadband radio and radar systems for telecommunications, industrial, automotive, railway, naval and defense markets. The Reicom competencies are:

- Software Defined Radio (SDR);
- Grid computing and cooperative computing;
- Audio and video streaming technologies;
- OFDM broadband radio;
- Ultra Wide Band (UWB) radio technology;
- Organic MESH Networks.

For the Railway market, the developed products are oriented to solve the problem of connectivity between the train and the ground and between coaches, and to bring real-time CCTV and Infotainment on board. Reicom developed the HSBRA™ (High Speed Broadband Radio Access) technology, relying on SDR for train to trackside broadband communications, tested over 300 km/h. HSBRA™ deals with the layer 1 and 2 of the OSI model, by implementing filtering, equalization, digital conditioning and management of the radio signal. HSBRA™ is designed to optimize and manage electromagnetic channel phenomena like fading, shadowing and Doppler effect. HSBRA™ is developed to be used in Vehicle to Vehicle (V2V) and Vehicle to Infrastructure (V2I) applications, addressing both safety and non safety critical applications. The technology relies on a Wi-Fi like technologies, dealing with IEEE 802.11 a/b/g/n and IEEE 802.11p standards, but also IEEE 802.20, IEEE 802.16 (WiMax) and 3G and LTE mobile networks.

Finally, some works are performed on millimeter-wave communication systems in order to solve the problem of degraded performance at high speed and growing traffic demand. In [85], the authors propose a Mobile Hotspot Network (MHN) of a millimeter-wave communication system as a mobile wireless backhaul. The solution proposes the physical layer design of the MHN uplink and downlink and the assessment of the performance of the two links.

Summary on Radio-Based Solutions

The section showed the different technologies that can be deployed as a dedicated track-to-wayside infrastructure, in order to provide broadband connectivity on board trains. The first observation is that there is just a few existing solutions. Most of them rely on WiMAX technology. However, as already mentioned, this kind of solutions allow an entire control of the Quality of Service, in terms of range and throughputs especially.

The Fluidmesh solution claims providing broadband connectivity up to 100 Mbps until 350 km/h. This solution is not yet largely deployed. To our knowledge, the first passenger service that aims to use it is the Amtrak trains.

The train-to-infrastructure solutions lead to high costs in terms of OPEX and CAPEX. A compromise has to be found between performance and costs (involved features are throughputs and coverage). This kind of solutions can be firstly used as a "gap-filler" solution.

2.3.2.2 Optical-Based Solutions

Optic Wireless Communications (OWC), also called Free Space Optics (FSO) represent an attractive technological solution in terms of throughput to obtain broadband Internet access on board trains. Indeed, FSO technologies offer large unregulated bandwidth allowing throughput up to Gbps, in addition to immunity to electromagnetic interferences and low Bit Error Rates (BER). Moreover, optical signals cannot penetrate walls and optical "print" being easily defined, transmission can be completely secured.

Studies in Japan

The Railway Technical Research Institute in Japan tested this technique, in collaboration with the Keio University [86–90]. Throughputs up to 700 Mbps are obtained at TCP level, for a speed of 130 km/h.

First works investigated a ground-to-train communication system using FSO technology [89]. Some BER experimental results using test train are given, showing that the proposed system is a promising candidate for train communication from the view point of BER characteristics. Works are then pursued by improving the system [87, 90]. Three different methods are tested:

- The leaky optical fiber method: this method requires installation of optical fiber along the track. It uses laser beam that flee through the fiber to establish a communication. The method allows to obtain continuous communication with the train;
- The "fan-shaped laser beam" method: this method uses a laser beam diffused with a concave lens. The lens radiates the laser beam in one horizontal direction. At reception, the laser beam is caught by the condenser lens. This is one of the characteristics of this method: laser transmitter can communicate with a wide area receiver;
- The "laser beam tracking" method: the transmitter consists of a laser transmission device and a mobile mirror. It transmits laser beams towards the receiver. This one is identified using an infrared beacon light. With the mobile mirror, the transmitter can follow the receiver and establish a continuous communication.

Preliminary tests were performed in order to compare the different methods. It follows that laser beam tracking method is the most efficient. It allows obtaining throughputs up to 400 Mbps (against 100 Mbps for the two other methods). Moreover, transmission distance is more important and dynamic mirror makes the solution much more flexible.

Authors detail the communication system by laser beam tracking adapted to the railway constraints. The communication device is embedded on board train (the mobile station) and its ground counterparts (base stations) send a laser signal in order to establish a bidirectional communication. Each of them transmits a light beacon signal, standing for an identifying signal, with a different wavelength from laser signal. To apply the laser beam tracking method to railway environment, it requires the deployment of many base stations in order to cover the entire railway line. Therefore, the system requires a handover mechanism between the different base stations. Problems to solve are then as following:

1. Connection has to be maintained whatever the speed of the train and the possible vibrations, for high speed. The mobile mirror has to operate in a very dynamical way;
2. Handover has to be performed rapidly and dynamically, even at high speed, connection being completely interrupted during the handover.

A developed prototype was able to record theoretical throughputs up to 1 Gbps. It consists of a mirror able to move in all 3D directions, which allows reducing size and weight of the device. Details of the development of the tracking mechanism and the optical equipment (lens, diodes and mobile mirror) are described in [88]. The minimization of the size and the weight of the lens is studied, in addition to the study of the beacon laser power and the types of lens to be used at reception (telephoto lens preferred to wide spectrum lens).

Handover mechanism between different base stations is also described. An optimized handover is implemented by improving standard protocol [86]. In this paper, measurement results in an emulated environment where a handover occurs every 5 s showed a packet loss rate of 2 % during the handover. The network is then divided in subnetworks because of its large size. Two types of handover have to be considered: the handover performed inside a same subnetwork, which is realized at layer 2 level, the link layer, and handover performed between two different subnetworks, realized at layer 3 level, the network layer. The system is based on the mobility protocol IPv6. Enhancements are performed from IPv6 at different steps of the handover, in order to minimize its duration.

Experiments are then realized in order to fix the "ideal" transmission distance depending on the number of base stations deployed along the track, which allow keeping a continuous communication during the entire trip. It follows that a distance between 300 and 400 m seems to be optimal.

Authors are interested in the influence of atmospheric conditions on the quality of the communication. The study is quite succinct and without numerical data. The given conclusion is that the quality of the communication depends on the visibility.

First tests of the entire system are set up. Initially, tests in static are performed. A first communication between two devices allows obtaining throughput at TCP level up to 923 Mbps. The transmission distance was tested until 360 m maintaining a communication. A glass was placed between two devices in order to simulate the train window. The transmission distance is reduced to 200 m, but without loss of throughput. A last test is implemented: a communication between a fixed base station and a mobile station put in a car moving at 100 km/h is realized; a maximal throughput of 656 Mbps is obtained.

After these preliminary tests, the system was tested on a train. Three bases stations are positioned along the track and connected to a control center. They are separated by 100 m from each other. At a speed of 130 km/h, throughputs between 500 and 700 Mbps are achieved. An important packet loss rate of about 20–30 % is observed, which represents a subject to improve. The handover time also remains high (about 0.4 s), which is due to the train vibrations causing instability of the infrared link. There is no significant observation regarding the influence of atmospheric conditions. A special effort is still to provide a protection of devices against condensation.

Finally, works conclude with HST trials on Shinkansen trains. The speed is about 240–270 km/h. A single base station could be installed. The handover mechanism could then not be tested. However, the communication between the mobile station on board train and the fixed base station was tested. The two stations could catch the beacon light for 0.7 s. A communication at PHY layer could be realized during 6 ms. However, no packets transmission could have been tested.

In [67], authors propose a collaboration of the Infrared Communication Device (IR-CD), presented above, and the LCX system deployed in the Shinkansen, presented earlier. The proposed system installs the IR at the upstream of the LCX system to keep the modification of the existing LCX system as small as possible. The LCX system and the IR system are not used at the same time; switch is performed when IR system is not available. The proposed system was implemented in Linux in order to evaluate the handover processing time (which is important in the case of the LCX system). It appears that handover time is short enough for passengers on Shinkansen (around 200 ms, near LCX handover time).

Studies in UK

Other works in UK were realized in order to evaluate performance of an OWC system to obtain a broadband Internet access on board trains. All the following papers are written by Paudel et al. from the Northumbria University of Newcastle.

In [91], the possibility to use MIMO optical systems is mentioned, in order to increase throughputs. However, the emphasis is placed on the description of a SISO optical system. Laser diodes are used and tests are performed on the size of the lens and the transmission power. The link budget of the developed system on short distances is computed. Nonetheless, no mobile experiments with higher distances (several hundreds of meters) were performed.

In [92], the communication protocol of the system is described. A base station is placed at a distance equivalent to two carriages of a train (around 42 m). Two receivers are placed for each carriage. Two types of applications are considered: the outdoor case and the indoor case (tunnel). In the first case, receivers are put on the roof of the train and base stations are deployed along the track, at the same height than receivers. In the second case, base stations are put on the ceiling of the tunnel, the configuration on train is the same. A system of control of switch on and off of the base stations with the passage of the train is presented. First tests on received power are realized. Optimizations on received power level with optical concentrators are presented.

In [93], the effect of turbulence on OWC is studied. Tests are performed in laboratory using an atmospheric chamber. The system is based on a transmission device using a LED, an infrared LED, an optical lens and a data source. The LED is modulated with a Non-Return-to-Zero (NRZ) and On-Off Keying (OOK) pseudo-random signal. Results show a high resistance of the developed system to the turbulence.

In [94], tests on throughputs depending on BER are performed in simulation, by varying the distance between two consecutive transmitters. The envisioned system is as follows. Transmitters consist of LED. They are put every 75 m on high voltage electric pylons located at about 1 m from the track. One transmit can cover three carriages (length of a carriage around 21 m). The receivers, consisting of photodetectors are positioned on the side or on the roof of the train at a height of 4 m approximately. A Lambertian model can represent this system. Simulations are performed with the Matlab tool in order to evaluate system performance. Thus, in order to obtain a BER of 10^{-6}, the optical coverage is around 75 m (3 carriages) for a throughput of 10 Mbps, 42 m (2 carriages) for a throughput of 100 Mbps, and 21 m (one carriage) for a throughput of 1 Gbps.

Finally, in [95], a model of the system specially designed for railway environment is presented. Laser are here used instead of LED, in order to obtain larger coverage area and more power. The system is then modeled by a Gaussian source, instead of a Lambertian source. A link budget analysis is performed showing a link margin of 17.75 dB for the worst atmospheric conditions. Simulation results with the Matlab tool are given in terms of BER performance of the system. It is showed that it is possible to have beam coverage up to 75 m for throughputs up to 50 Mbps.

Summary on Optical-Based Solutions

Works presented in this section highlighted different aspects. First of all, it appears that works performed in UK are largely dominated by simulations and no measurement in real sites, with railway constraints were performed yet. Conversely, works in Japan are quite advanced and promising for a new option for providing Internet on board trains. Very high throughputs can be obtained at very high speeds. Throughputs up to several hundred of Mbps were measured in a real site. However, the major drawback remains the cost of CAPEX and OPEX of optical-based solutions. Optical terminals have to be deployed at least every 400 m along the track, which leads to a very high investment. Furthermore, this rolling out leads to high cost of maintenance. Finally, optical solutions performance are very dependent on atmospheric conditions.

2.4 Summary on How to Provide Broadband Internet on Board Trains

2.4.1 General Remarks

There are several technological solutions that can be used to provide a broadband Internet access on board trains. The list of solutions presented in this chapter is not exhaustive, due to the constant evolution of the subject. Many new solutions are emerging regularly. Furthermore, each solution has its own advantages and drawbacks that will be summarized in this section. It appears that whatever the solution used, similar conclusions can be highlighted:

- Several Railway Companies establish a quota system on the used bandwidth in order to limit the required throughputs. For instance, Amtrak has implemented a rate limiting on all East coast and Midwest services in March 2014: passengers are allowed to consume up to 250 MB of date. Once exceeded, their data transfer rate is limited to 200 kbps to reduce data consumption [59]. Such a quota system is also used in Sweden and in the Netherlands, by limiting the speed per use to 150 kbps for the latter;
- A majority of the solutions were first rolled out in the 2000s. For the most part, they were, or are currently, upgraded, especially with the possible use of the Ka band for satellite solutions, and the deployment of the 4G cellular technology in the different countries for the cellular solutions.

2.4.2 Comparison of the Different Technologies

Table 2.4 presents a comparison and a summary on the different technologies, in terms of throughputs, latency and specific advantages and drawbacks. The table highlights that satellite solutions offer the advantage to be able to use the existing infrastructure. However, throughputs are limited and a "gap-filler" is necessary in presence of obstacles, such as tunnels.

Cellular-based solutions have the main advantage of having no infrastructure to deploy, by aggregating several public MNOs. However, it is not possible to manage the Quality of Service, in terms of throughputs, coverage, latency, etc.

All the solutions requiring a dedicated infrastructure have the main drawback to lead to high cost in terms of CAPEX and OPEX. However, Quality of Service can be managed and performance are quite better. First solution is to install radio terminals, relying on Wi-Fi or WiMAX technologies. Throughputs up to 100 Mbps can be obtained. The second solution consists in using fiber optic to connect the different base stations. As presented in the dedicated section, this system relies on low cost base stations allowing a reduction of handover times, a reduction of costs and an increase of performance in terms of throughputs. Finally, the third solution, the optical-based solutions, allows clearly obtaining the best performance in terms of throughput and latency. Nevertheless, it requires a heavy installation along the track (optical terminals put every 400 m), which leads to very high CAPEX and OPEX. Furthermore, optical-based solutions are dependent on atmospheric conditions, such as now, rain and fog.

Next section is dedicated to the broadband wireless communications that can be used for operational needs. Indeed, providing broadband links between train and

Table 2.4 Summary of the different technologies to provide internet on board trains

	Satellite	Cellular	Radio terminals	RoF	Optical
Throughput	>10 Mbps	>10 Mbps	>100 Mbps	/	>10 Gbps
Latency (ms)	>400	>200	>100	>100	>50
Advantages	Existing infrastructure	No infrastructure, low cost	Average throughput, seamless communications	Low cost base stations, seamless communications	Very high throughput, seamless communications
Drawbacks	Limited throughput, communication failures due to obstacles (tunnels, relief, etc.),	Possible limited coverage, limited throughput	High costs	High costs	Heavy infrastructure needed, influence of atmospheric conditions, very high costs

ground for Internet on board allows the railway operators to perform other operational applications requiring high throughputs, such as maintenance and video surveillance.

2.5 Broadband Wireless Communications for Operational Needs

Broadband wireless communications needs for railway applications are quite growing since several years, as presented in Chap. 1. Services to passengers are very demanding in terms of throughputs, as presented above. Other applications requiring broadband wireless communications are presented in this section. Among these applications, we studied maintenance activities, video surveillance applications and other applications, such as smart metering and freight. We decided to separate the maintenance and the video surveillance applications even if they are strongly related. On one hand, maintenance can be performed by using video techniques, but also other techniques, such as WSNs. On the other hand, the video surveillance can be used for maintenance purposes but also for other applications, such as security.

2.5.1 Maintenance

With the rapid innovations in computer science, an increase of amount of data, coming from many sources is observed in many domains, such as railway domain [96]. The traditional data-collecting methods rely on selected measurements over specific assets. Henceforth, it is now possible to perform continuous collection of information from several sources from the entire railway system. This phenomenon leads to an improvement of monitoring and maintaining of railway system by using real-time information.

Maintenance operations lead to accumulative delays, that can disturb railway traffic. Scheduling maintenance work is quite difficult due to the high occupancy. There is then a high demand for efficient and reliable maintenance operations based on frequent measurements of the different parts of the railway system. Thus, continuous data processing and high quality decision making are required. Video is one of the technology that can be used to monitor the systems leading to high required throughputs to transmit the data.

ERRAC targeted for the year 2020 to double passenger and freight traffic by rail. Taking into account this expected growth in passenger and freight volumes and the aging of existing infrastructure, maintenance needs and costs are likely to increase significantly in the coming years. It represents then a major objective to reduce the life-cycle cost of the railway transport. One way to reduce the cost of rail services is to limit expenditures linked to the operation of services, including energy consumption and maintenance.

Maintenance of rail network infrastructure has to be safe, reliable, cost-effective and sustainable. A significant part of the costs for reliable high capacity infrastructures is related to intensive maintenance, most of which is preventive. Thus, maintenance costs have to be reduced by especially simplifying procedures and automation. Better maintenance strategies can be based on risk-based or condition-based analytics, using more reliable sensor technology to detect real asset condition. Furthermore, enhanced maintenance procedures can be based on remote infrastructure condition monitoring and automated, self-inspecting, adjusting and correcting devices. The maintenance systems can be built on cutting edge measurement and monitoring tools that provide static and dynamic data from all relevant components of the rail infrastructure, using train-borne, wayside and remote sensing measurement and monitoring systems. Automation should be achieved for routine maintenance checks, as well as for repetitive tasks, such as track relaying, ballast renewal, tamping and alignment.

The amount of data that has to be raised from train to ground are then more and more important and requires high capacity wireless communications between train and wayside. An alternative could be that some pre-analysis on board the train is performed to reduce train-to-ground transfer. All the systems are presented in this section.

2.5.1.1 Definitions Related to Maintenance Systems

Three main types of maintenance can be identified [97]:

- The corrective maintenance: according to the definition of the standard NF EN 13306, the corrective maintenance is carried out after failure detection and aims to restore an asset to a condition in which it can perform its intended function. The corrective maintenance can be divided into two cases: the "immediate corrective maintenance" based on an immediate intervention after a failure, and the "deferred corrective maintenance" where the work is delayed in conformance to a given set of maintenance rules;
- The preventive maintenance: it aims to maintain equipment and facilities in satisfactory operating condition by providing inspection, detection and correction of failures, either before they occur, or before they develop into major defects. It can also represent the tests, measurements, adjustments, and parts replacement performed specifically to prevent faults from occurring. Two preventive maintenance can be identified: the planned maintenance and the condition-based maintenance. The two subgroups differ regarding the determination of maintenance time, or the determination of moment when the maintenance should be performed;
- The predictive maintenance: it goes further the preventive maintenance by predicting the moment the maintenance should be performed. The predictive maintenance allows reducing cost over routine or time-based preventive maintenance. Indeed, maintenance tasks are performed only when warranted. The predictive maintenance allows convenient scheduling of corrective maintenance and prevents unex-

pected equipment failures. These techniques lead to shorter and fewer "planned stops" and then to an increase of availability. Furthermore, an increase of equipment lifetime, plant safety and a decrease of accidents with negative impact on environment are the consequences of the use of predictive maintenance.

2.5.1.2 Optimization Models for Maintenance

Several works deal with the optimization of the railway maintenance system. Some models are developed to optimize maintenance scheduling and perform preventive works.

In [98], an optimization model is developed to improve rail maintenance decisions, relying on a dynamic schedule of preventive maintenance activities. Maintenance works are assigned to different time periods and different track segments. Approximation methods are introduced to deal with the large amount of instances and provide the best possible solutions.

In [99], different approaches are proposed to deal with the problems of the railway preventive maintenance scheduling. Different algorithms and techniques are tested, such as an hybrid genetic algorithm, ontology-based modeling. The paper also examines the different strategies applied all over the world for solving the maintenance scheduling problem. In US, the aim is to minimize the overall cost of all maintenance jobs while in Europe, the reduction of immobilization of the trains is a main concern.

In [100], a Markov technique is used to model a section of railway track in UK and aims to be extended to implement a global strategy of maintenance decision process for railway track.

In [101], an optimal maintenance model combined with maintenance activity is proposed for High Speed Railway Signaling System in China. In [102], the authors deal with predictive maintenance and management for Indian Railways, by introducing techniques of mining data, such as clustering approach.

In [103], the authors describe the works performed in the SURFER project (2010–2013) dealing with a discriminative model for online predictive diagnosis of train door system. The project proposed a system of diagnosis and detection of embedded failures to develop a predictive maintenance and increase availability of equipment. According to Bombardier, a partner of the project, the doors access represent 5 % of the cost and 30 % of failures in the case of Francilien. It is then really interesting and cost-effective to intervene before the failures occur. The data of selected indicators are transmitted in real-time to a distant terminal via 3G communication links.

2.5.1.3 Track Maintenance

Many works deals with the techniques dedicated to track maintenance. Rail inspection is a quite important task in railway maintenance domain. Inspection can be performed manually by human operator walking along the track and searching for visual anomalies. However, such method is slow and lack of objectivity, depending on the observation of the person in charged. Other automatic and intelligent system are developed.

In [104], the authors present a real-time Visual Inspection System for Railway (VISyR) maintenance, able to detect automatically the fastening bolts that fix the rails to the sleepers. Images acquisition is performed with a digital line-scan camera. In [105], a computer vision system is implemented to improve the track maintenance. The system relies on video acquisition and analysis. Algorithms are developed to perform detection, segmentation and defect assessment of track components. In [106], a video monitoring system is developed for high speed railway applications, relying on three parts. The first part is composed of ground condition collectors installed on track to collect railway conditions, encode, encapsulate and transmit information to wireless relay network through fiber cables. The second part of the system deal with the wireless relay network, that transmit information from track to vehicle. The third part consists of vehicles, that decode information and show the conditions to the driver. The objective is to provide a view for the driver beyond his LOS. Simulations are performed to evaluate the theoretical performance of the system in terms of throughput, outage probability and average delay.

In order to increase track lifetime, systems and methods still have to be developed and enhanced for measuring stress, degradation, stiffness, friction, defects or impacts from climatic changes on tracks. At the same time, damage prevention strategies can also be enhanced, using integrated health monitoring systems and innovative methods for on-site rail manipulation [107]. Thus, many studied systems rely on video acquisition and transmission to optimize the maintenance, such as the automatic track monitoring. Broadband wireless links are then required between train and ground to transmit the large amount of video data. Furthermore, the rolling stock can be used to monitor the track or install dedicated system on the track. The rolling stock already allows to monitor the track by analysing the accelerometer default and traction default for instance.

2.5.1.4 Cloud Computing and Big Data

The management of the huge amount of data required for the different railway applications, such as maintenance, can be performed by the new techniques of cloud computing and Big Data. Cloud computing is a computing term relying on the deployment of groups of remote servers and software networks to centralize data storage and online access to computer services or resources. Cloud computing aims to share resources over a network. The term “moving to cloud” can refer to an organization moving away from a traditional CAPEX model, relying on dedicated hardware with depreciation of it over a period of time, to the OPEX model, relying on the use of shared resources with a pay-per-use system. Cloud computing can be divided into three levels of services:

1. Software as a Service (SaaS): users can have access to application software and databases over Internet, without required installation of software. The cloud infrastructure and platform are not managed by the users in this case;
2. Platform as a Service (PaaS): users can have access to a computing platform, including typically operating system, programming language execution environment, database, and web server;
3. Infrastructure as a Service (IaaS): users can have access to a remote computer. The user can then install the needed operating system and put his applications.

Big Data is a broad term for data sets very large or complex, for which traditional data processing can not be applied. The challenges can then include analysis, capture, curation, search, sharing, storage, transfer, visualization, and information privacy. Big data can be described by the following characteristics (5Vs):

- Volume: the quantity of generated data is very important. The size of the data determines the value and the potential of the data under consideration and if it can be considered as Big Data;
- Variety: the category to which Big Data belongs is also a very essential point needed to be known for the data analysis;
- Velocity: it represents the speed of generation and processing of data;
- Variability: it refers to the inconsistency which can be shown by the data at times and leads to difficulty to handle and manage the data effectively;
- Veracity: accuracy of data analysis depends on the veracity of the source data, knowing that the quality of the data can strongly vary;
- Complexity: data management can become a very complex process, especially when large volumes of data come from multiple sources. Data need to be linked, connected and correlated in order to be able to grasp the information that is supposed to be conveyed by these data. This situation is therefore termed as the "complexity" of Big Data.

In [108], cloud computing concept is used to perform remote maintenance, such as remote switch and crossing monitoring, relying on the SURVAIG application specifically developed. Remote monitoring equipment combined with PC server rent on the Internet (IaaS) are used. All the applications are then installed on one SURVAIG platform (PaaS). The SURVAIG service is then accessible via the Internet (SaaS). The system is installed in parallel of the existing infrastructure. It performs data measurements and also analysis of data: Failure Mode, Effects and Criticality Analysis (FMECA). The SURVAIG application can then be rent monthly. The infrastructure managers can reduce their investments with the absence of the initial cost and the maintenance cost of a server. Thus, the SURVAIG system acquires wide range of data in real time, processes then locally, archives them in a deported station and performs immediate or deferred analysis, to predict maintenance works and anticipate failures or interventions. For railway applications, devices on track are connected to the cloud via local deployed cellular networks, such as 3G or 4G.

In [96], the authors study the possibility to use Big Data techniques to enhance maintenance decisions for railway tracks. The conditions of the 5 Vs of Big Data are discussed for railway monitoring systems. Indeed, several problems have to be overcome to collect and store the data. Efficient methods have to be implemented to analyze the Big Data for decision making. The authors present a case study of an embedded monitoring solution relying on Axle Box Acceleration (ABA) measurements, GPS and video of the track. Existing ABA systems can be based on ultrasonic and eddy current techniques or on video from cameras mounted below the train. It is showed that Big Data techniques present a great potential to enhance maintenance decisions. In the case study in Dutch railways, 1 terabyte of raw data per day have to be managed. Selective data processing is implemented, it is demonstrated that all parts of the track can be monitored.

Cloud computing and Big Data can thereby represent quite relevant means to deal with the huge and growing amount of data circulating between train and ground for different applications, such as maintenance and monitoring works. The digital transformation must come with the system evolution.

2.5.1.5 Wireless Sensor Networks (WSNs)

Wireless Sensor Networks (WSNs) and Internet of Things (IoT) are largely explored in many domains, relying on a large number of physical objects being connected to each other and/or to the Internet [109, 110].

In [111], it is showed that research in the field of WSNs is very dynamic, and there are high expectations regarding applications of sensor networks. A state of the art on recent developments in WSN technologies and their applications is performed. The obstacles in the application of WSN that should be addressed in order to push the technology further are identified. A taxonomy of WSN applications is presented, highlighting the particular case of ITS.

WSNs are particularly studied to perform railway monitoring. Distributed sensor technologies can be used to perform structural health monitoring of tracks, carriages and other equipment in the railway system. In recent years, sensing technologies grew up and a large range of sensors became cheaper. Furthermore, the cost of using public network has decreased significantly. Machine-to-Machine (M2M) SIM cards will then expand largely. It is less and less necessary to build dedicated private networks which represents a high sustaining cost. This phenomena lead to a rapid expansion of WSN systems. WSNs can be implemented for maintenance applications, such as condition monitoring of railway systems. WSNs rely on the deployment of several sensors and on networking technologies to couple the different sensors. Thus, recent advances in wireless telecommunications and adhoc networking also enable the development of these technologies. WSNs can be used for monitoring the railway infrastructure (bridges, rail tracks) and also perform vehicle health monitoring (chassis, bogies, wheels).

In [112], a survey of WSN systems for monitoring in the railway industry is performed. The paper deals with the engineering solutions developed, such as the types and different uses of sensor devices, and the identification of sensor configurations and networks. In [113], a monitoring system is developed to detect rail damage.

An integration of three different methods is realized: an optical fiber grating method, an optical imaging method, and a Lamb guided wave method.

Studies on decentralized control of remote trackside objects, such as level crossing are considered without using trackside cables [114]. To do this, locally-derived power supply and safe and secure radio communications have to be set up. We talk about "connected objects". Sensors capabilities relying on self-adjustment can also be used for self-diagnosis and remote condition monitoring leading to a reduction of failure and maintenance costs for Switch and Crossing applications. Finally, Railway Integrated Measuring and Monitoring System (RIMMS) are being considered to measure relevant data, monitor the status of railways critical assets and then enhance the monitoring and maintenance applications [114].

The section presented the large domain of applications regarding the railway maintenance, diagnosis and monitoring. More and more data transit between train and ground because of the possibility to perform continuous collection of information from several sources from the entire railway system. All the collected data can then be used to optimize maintenance works by using real-time information. The interest of real time is mainly for decision making, it can then confirm a train mission or route the train towards a maintenance depot. The life-cycle cost of the global railway system can then be minimized because of the reduction of the cost of maintenance.

2.5.2 *Video Surveillance*

Nowadays, video surveillance systems are largely deployed in public spaces, including the transport domain. Many researches and studies are conducted in this thematic to help human operator in charge of analyzing the recorded images. The earliest systems were called Closed-Circuit TeleVision (CCTV) because the data were not transmitted outside of the environment being monitored. Currently, systems rely on IP cameras linked by an Ethernet network. The objective is to perform maximum interconnection and interoperability of all the video surveillance systems inside a specific environment, such as public places or transport environment. Strong constraints are required in terms of integration, maintenance and communication. Reliable wireless broadband communication systems have to be implemented to link the video surveillance systems on board vehicles and on the ground. Video surveillance can be used for surveillance of railway infrastructures (tracks, terminals and stations) or for surveillance on board vehicles [115].

2.5.2.1 Video Surveillance of Railway Infrastructures

Several projects were dedicated to surveillance of railway infrastructures. The first video devices were entirely managed manually. However, human operator can have difficulty to focus on a large number of video screens at once and over a long period of time. The CROMATICA and PRIMATICA research projects were the first collaborative projects dealing with the development of tools to aid the exploitation of video surveillance systems [115]. The projects elaborated a breakdown of the requirements in terms of automatic functions that systems of video surveillance in metro stations, train stations and airports could offer. The technical feasibility was demonstrated for some specific user cases, such as estimation of the density of passengers and crowd detection, detection of fallen objects on tracks, etc. The different user cases were then tested and a decentralized processing at camera level was implemented, in order to transform the centralized system to a distributed system, able to react more efficiently to rapid events. The CARETAKER project pursued the works by exploiting audio and video streams in metro stations. More recently, the VANAHEIM project worked on the automatic analysis of video and audio streams in a metro station.

The PANSAFER project aimed to improve the safety at level crossings. The project defined and performed a functional analysis of the different typical accident scenarios. The objective of the project was also to study the feasibility of an automatic system able to detect and alert road users in case of an accident scenario, relying on video perception tools and wireless communications, from IEEE 802.11p standard, which adds Wireless Access in Vehicular Environments (WAVE). The objective is to send the information of an abnormal situation occurring to the users, so they adapt their behavior to solve the incident. Other works can be also cited on the surveillance of level crossings, relying on video [116], but also Ultra Wide Band (UWB) technologies [117].

Video surveillance for railway infrastructure monitoring, such as railway stations and level crossings, does not require broadband wireless communications from train to ground. The following section deals with this subject.

2.5.2.2 Video Surveillance on Board Vehicles

Video surveillance systems were embedded in vehicles since the early 2000s in France to ensure the safety of users throughout their journey, including in a multimodal context (use of different modes of transport). Some research projects, such as TESS and EVAS projects, dealt with the video surveillance of buses. The objective of the TESS project was to develop new information and safety services, relying on satellite geolocalization and on the coupling of audio and video by equipping the bus of several cameras and microphones. The system allows then to compensate the difficulties encountered by the image interpretation methods. Works were pursued in the EVAS project, that developed a real-time system using smart audio and video surveillance and a wireless communication between the bus and the control center, relying on a MIMO-WiMAX wireless link.

In the railway domain, transport operators expressed the need of video surveillance systems, noting the lack of surveillance solution particularly because of the absence of efficient means of transmission between the trains and the control center. The two main axes of research in the railway domain are then to develop automatic audio and video systems and to implement broadband wireless communication systems between train and infrastructure. Several research projects worked on these topics. The SAMSIT project initiated the works in the context of railway environment.

The BOSS project [118] developed a system able to interconnect all the actors involved in the detection of abnormal events, such as the audio and video surveillance system, the conductors, the driver of the train and the control center. A communication system was developed, based on an IP gateway able to communicate both within the train itself and with the outside world, adapting the throughputs and the quality of the connection, and managing mobility of information both on board and to the ground. The system was tested in a Madrid suburban train. Finally, the SURTRAIN project allowed the integration of an audio and video surveillance system, designed in accordance to railway norms, in a railway vehicle. It was installed as a permanent fixture inside the vehicle.

Few works deal with the video surveillance in the railway domain. A lot of CCTV systems are deployed in metro, where wireless Wi-Fi like networks are easily deployed in the particular tunnel environment. In the case of conventional trains, such as suburban trains, few systems were studied. Finally, and to our knowledge, no systems are installed in the case of HST due to the absence of efficient wireless transmission link between train and ground.

2.5.3 *Other Operational Applications*

2.5.3.1 Smart Metering

One of the keys of innovation for the future of the railway is the enhancement of the energy efficiency. This objective can be reached by using methods of smart metering. Smart metering relies on a distributed energy resource management system, which aims to manage the different energy flows of the entire railway system [107]. A smart metering system allows to obtain a knowledge of all consumers and generators energy flows, which enable to set up energy savings, losses reduction and efficient energy management.

The WP11 of the In2Rail European project [119] aims to develop a Smart Metering for a Railway Distributed Energy Resource Management System (RDERMS). The main objectives of the works are to design an open system dedicated to the fine mapping of different energy flows within the whole Railway System on a synchronized time basis. The workflow methodology is defined around the design of the physical and software support, linking together measurements taken at different locations, subsystems, and temporal information. The smart metering aims to reduce the energy bill, optimise asset management and increase capacity. Such systems rely on sensors deployed in the railway system (on board and at trackside), communication between sensors and communication between the train and the ground.

Smart metering systems relies then in particular on sensor networks and communication protocols. Intelligent WSN have to be set up to guarantee successful collection of distributed sensor measurements. Then, trackside and on board devices can be installed. Wireless communication systems have then to be set up from train to ground to transmit the amount of different energy measurement data, requiring broadband links.

2.5.3.2 Freight

Rail freight generates low level of external costs and allows reducing the environmental impact. Indeed, rail is the most eco-friendly land transport mode for freight with much more energy consumption and CO2 emissions than road freight or transport by waterways. The key challenge for rail freight is then to offer an attractive, reliable, fast and cost-efficient alternative to road freight [107].

Rail freight suffers of a stagnation in the freight domain due to legal barriers and operational and technical problems, which impact the overall capacity and performance of the rail freight. The cost competitiveness and the reliability of freight services need then to be improved.

Two main challenges can then be identified:

- Acquisition of a new service-oriented profile for rail freight services, relying on on-time delivery and competitive prices;
- Increase of productivity by addressing current issues, such as interoperability and development of cost-effective solutions, by optimizing existing infrastructure and promoting transfer from other sectors to rail freight.

Freight trains has then to follow technological evolutions to improve operational performance, interoperability and increase capacity, such as [114]:

- Increase automation of marshalling yards and then reduce train setup;
- Improve dynamic train performances;
- Provide real time information, such as health monitoring, control and monitoring of dangerous goods;
- Enable interaction and exchange of information from train to ground.

Freight wagons has then to be equipped with mechatronic system with sensors, data processing and communication systems. In [120], The French Train-MD research project is presented, dealing with the transport of hazardous goods. The project aimed to design and develop an innovative system to better manage the hazardous goods traffic, relying on tracing facilities, such as GPS, GSM or GPRS balises. Remote real-time diagnosis is also performed with sensors embedded in the wagons. Finally in [121], wireless monitoring is performed in order to improve the safety of freight trains.

This section presented the large number of operational applications requiring broadband communication links between train and ground. These applications are more and more demanding in terms of throughputs, due to the quite growing data flow coming from the entire railway system. The next chapter is dedicated to the future challenges and opportunities of railway and points out the different emerging technologies and future trends in terms of railway communications.

References

1. Juan Pablo Conti. Hot Spots on Rails. *Communications Engineer*, 3(5):18–21, November 2005.
2. Daniel T Fokum and Victor S Frost. A survey on methods for broadband internet access on trains. *Communications Surveys & Tutorials, IEEE*, 12(2):171–185, 2010.
3. G. Bianchi, N. Blefari-melazzi, E. Grazioni, S. Salsano, and V. Sangregorio. Internet access on fast trains: 802.11-based on-board wireless distribution network alternatives. In *IST Mobile & Wireless Communications Summit*, pages 15–18, Aveiro, Portugal, June 2003.
4. Christopher J Hansen. WiGiG: Multi-gigabit wireless communications in the 60 GHz band. *Wireless Communications, IEEE*, 18(6):6–7, 2011.
5. Eldad Perahia, Carlos Cordeiro, Minyoung Park, and L. Lily Yang. IEEE 802.11ad: Defining the Next Generation Multi-Gbps Wi-Fi. In *7th IEEE Consumer Communications and Networking Conference*, pages 1–5, Las Vegas, January 2010. IEEE.
6. Steven J. Vaughan-Nichols. Gigabit Wi-Fi Is on Its Way. *Computer*, 43(11):11–14, November 2010.
7. IEEE Std 802.11ad-2012 (Amendment to IEEE Std 802.). *IEEE Standard for Information technology–Telecommunications and information exchange between systems–Local and metropolitan area networks–Specific requirements-Part 11: Wireless LAN Medium Access Control (MAC) and Physical Layer (PHY) Specifications Am*, pages 1–628, 2012.
8. Atchareeya Kongbuntad, Monthippa Uthansakul, and Peerapong Uthansakul. Improved beamforming code book for WiGig using Maximal Ratio Combining. In *The International Conference on Information Networking (ICOIN)*, pages 112–115, Phuket, Thailand, February 2014. IEEE.
9. Aimeric Bisognin, Diane Titz, Fabien Ferrero, Romain Pilard, Carlos A. Fernandes, Jorge R. Costa, Christian Corre, Pierino Calascibetta, Jean-Michel Rivire, Alexis Poulain, Christian Badard, Frric Gianesello, Cyril Luxey, Pierre Busson, Daniel Gloria, and Didier Belot. 3D printed plastic 60 GHz lens: Enabling innovative millimeter wave antenna solution and system. In *IEEE MTT-S International Microwave Symposium (IMS)*, pages 1–4, Tampa, Fl, June 2014.
10. C. H. Chan, K. B. Ng, D. Wang, H. Wong, and S.-W. Qu. Antennas for 60 GHz high-speed radio systems. In *IEEE International Workshop on Electromagnetics; Applications and Student Innovation (iWEM)*, pages 1–2, Chengdun China, August 2012.
11. E. Juntunen, A. Tomkins, A. Poon, J. Pham, A. El-Gabaly, M. Fakharzadeh, H. Tawfik, Yat-Loong To, M. Tazlauanu, B. Lynch, and R. Glibbery. A compact antenna-in-package 60-GHz SiGe BiCMOS radio. In *IEEE Radio Frequency Integrated Circuits Symposium*, pages 287–288, Tampa, FL, June 2014.
12. H Mopidevi, HV Hunerli, Engin Cagatay, N Biyikli, Marc Imbert, Jordi Romeu, Lluis Jofre, and Bedri Artug Cetiner. Three-Dimensional microfabricated broadband patch antenna for WiGig applications. *Antennas and Wireless Propagation Letters, IEEE*, 13:828–831, 2014.
13. Kung Bo Ng and Chi Hou Chan. A differentially-fed complementary antenna for WiGig applications. In *Asia-Pacific Microwave Conference Proceedings (APMC)*, pages 313–315, Seoul, Korea, November 2013.
14. R Pilard, F. Gianesello, and D. Gloria. 60 GHz antennas and module development for WiGig applications: Mm-wave antenna-systems convened session. In *6th European Conference on Antennas and Propagation (EUCAP)*, pages 2595–2598, Praha, March 2012.
15. B. Razavi, Z. Soe, A. Tham, J. Chen, D. Dai, M. Lu, A. Khalil, H. Ma, I. Lakkis, and H. Law. A low-power 60-GHz CMOS transceiver for WiGig applications. In *Symposium on VLSI Circuits (VLSIC)*, Kyoto, June 2013.
16. Nobuo Saito, Takayuki Tsukizawa, Naganori Shirakata, Takahito Morita, Kiyoshi Tanaka, Jun Sato, Yu Morishita, Masaki Kanemaru, Ryo Kitamura, Tal Shima, et al. A fully integrated 60-GHz CMOS transceiver chipset based on WiGig/IEEE 802.11 ad with built-in self calibration for mobile usage. *Solid-State Circuits, IEEE Journal of*, 48(12):3146–3159, 2013.

17. T. Tsukizawa, N. Shirakata, T. Morita, K. Tanaka, J. Sato, Y. Morishita, M. Kanemaru, R. Kitamura, T. Shima, T. Nakatani, K. Miyanaga, T. Urushihara, H. Yoshikawa, T. Sakamoto, H. Motozuka, Y. Shirakawa, N. Yosoku, A. Yamamoto, R. Shiozaki, and N. Saito. A fully integrated 60GHz CMOS transceiver chipset based on WiGig/IEEE802.11ad with built-in self calibration for mobile applications. In *IEEE International Solid-State Circuits Conference Digest of Technical Papers (ISSCC)*, pages 230–231, San Francisco, February 2013.
18. Sungyun Jun and Kai Chang. A 60 GHz monopole antenna with slot defected ground structure for wigig applications. In *IEEE Antennas and Propagation Society International Symposium (APSURSI)*, pages 2139–2140, Orlando, July 2013.
19. http://blog.lifi-led.com/.
20. http://www.lifinews.com/.
21. H. Ghannoum. Li-Fi project at sncf. In *Forum NTIC IFSTTAR*, 2013.
22. Hassan Ghannoum and David Sanz. Internet Onboard: technical analysis. In *5th International Workshop on Communication Technologies for Vehicles, Nets4cars/Nets4trains*, Lille, France, May 2013.
23. Juan-Carlos Maureira. *Internet on Rails*. Phd thesis, University of Nice - Sophia-Antipolis, January 2011.
24. Paolo Chini, Giovanni Giambene, and Sastri Kota. A survey on mobile satellite systems. *International Journal of Satellite Communications and Networking*, 28(1):29–57, 2010.
25. Yurong Hu and Victor OK Li. Satellite-based internet: a tutorial. *IEEE Communications Magazine*, 39(3):154–162, 2001.
26. Andreas Fabri, Txomin Nieva, and Paolo Umiliachi. Use of the Internet for Remote Train Monitoring and Control: the ROSIN Project. In *Rail Technology '99*, London, UK, September 1999.
27. Andrea Gatti. Trains as Mobile devices: the TrainCom project. In *Wireless Design Conference*, London, UK, May 2002.
28. V. Schena and F. Ceprani. FIFTH Project solutions demonstrating new satellite broadband communication system for high speed train. In *IEEE 59th Vehicular Technology Conference. VTC-Spring*, volume 5, pages 2831–2835. IEEE, May 2004.
29. http://www.traincom.org/.
30. Jerome Billion and Didier Van den Abeele. ICOM: A Communication Framework for Interoperable European Railways. In *7th International Conference on Intelligent Transportation Systems Telecommunications (ITST)*, pages 1–6, Nice, France, June 2007. IEEE.
31. R. Shingler, G. Fadin, and P. Umiliacchi. From RCM to predictive maintenance: The InteGRail approach. In *4th IET International Conference on Railway Condition Monitoring*, pages 1–5, June 2008.
32. M. Emmelmann. TCP/IP over Satellite. Technical report, Internal working document, Fraunhofer FOKUS.cats, Berlin, Germany, July 2000.
33. Marion Berbineau, Maria Chennaoui, Christophe Gransart, Hossam Afifi, Jean-Marie Bonnin, David Sanz, Rmy Coutureau, and David Duchange. High Data Rate Transmissions for High Speed Trains, Dream or Reality? Technical state of the art and user requirements, Train-IPSAT Project - WP1. Technical report, Synthesis INRETS N51, ISBN 278-2-85782-643-5, ISSN 0769-0274, June 2006.
34. http://www.indracompany.com/en.
35. http://www.icomera.com.
36. David Wilson. 10 Years of Wi-Fi on the East Coast Mainline. In *The WiFi on Trains Conference - Train Communications Systems*, London, UK, June 2014.
37. SCI Verkehr. Railway Wi-Fi Applications - Current Projects and Global Developments. Technical report, 2010.
38. 21Net. Broadband to Trains. In *ARTES Workshop*, Rome, Italy, April 2013.
39. Henry Hyde Thomson. Challenges of Internet & Multimedia on Trains. In *The WiFi on Trains Conference - Train Communications Systems*, London, UK, June 2014.
40. NTV. NTV Presentation. In *The WiFi on Trains Conference - Train Communications Systems*, London, UK, June 2014.

41. http://www.21net.com.
42. http://www.latribune.fr/entreprises-finance/services/transport-logistique/20141105trib5db4a0ea2/thalys-va-installer-un-wi-fi-plus-rapide-dans-ses-trains-sncf.html.
43. Dave Niravkuamr. 2nd Generation Of Secured Satellite Based Broadband System for efficient train operation and passenger Wi-Fi. In *The WiFi on Trains Conference - Train Communications Systems*, London, UK, June 2014.
44. Gilat. Connectivity for Trains. Technical report, June 2011.
45. Gilat. Bringing Broadband Connectivity to Trains with Satellite Communication. Technical report, 2013.
46. David Sanz. Satellite Technologies for Broadband Internet Access Onboard High Speed Trains. In *7th World Congress on Railway Research*, Montreal, Canada, June 2006.
47. David Sanz, P. Pasquet, P. Mercier, Bernadette Villeforceix, and David Duchange. TGV Communicant Research Program: from research to industrialization of onboard, broadband Internet services for high-speed trains. In *8th World Congress on Railway Research*, Seoul, Korea, May 2008.
48. B. Villeforceix. Delivering Broadband Internet Access for High Speed Trains Passengers Using an Innovative Network Mobility Solution. In *Nets4cars Nets4Trains*, pages 34–44, 2011.
49. Panagiotis Georgopoulos, Jose Andre Moura, Rafidah Md. Noor, Ben McCarthy, and Christopher Edwards. Theoretical and Practical Survey of Backhaul Connectivity Options. Technical report, Lancaster University, Computing Department, June 2010.
50. Keon Jang, Mongnam Han, and Soohyun Cho. 3G and 3.5G Wireless Network Performance Measured from Moving Cars and High-Speed Trains. In *1st Workshop on Mobile Internet through cellular networks (MICNET)*, pages 19–24, September 2009.
51. http://nomad-digital.com/.
52. Robert Oerton. WiFi on East Midlands Trains. In *The WiFi on Trains Conference - Train Communications Systems*, London, UK, June 2014.
53. Lesley Brown. Eurostar brings Wi-Fi on board. *EURAILmag*, 26:189–190, 2014.
54. http://www.viarail.ca/en/about-via-rail/capital-investment/article/worlds-fastest-wifi-rails.
55. Andres Arjona, Antti Ylski, and Jussi Kerttula. Live Network Performance Challenge: FLASH-OFDM vs. HSDPA. In *22nd International Conference on Advanced Information Networking and Applications (AINA)*, pages 918–925, March 2008.
56. Emlyn Clment. First Taste of Wi-Fi on Budapest-Sopron. *EURAILmag*, 24:120–121, 2011.
57. Justin Stenner. Leveraging Wi-Fi: Current & Future Use at Heathrow Express. In *The WiFi on Trains Conference - Train Communications Systems*, London, UK, June 2014.
58. Bruce Chow, Ming-Li Yee, Michael Sauer, and Anthony Ng'Oma. Radio-over-Fiber Distributed Antenna System for WiMAX Bullet Train Field Trial. In *IEEE Mobile WiMAX Symposium*, pages 98–101, Napa Valley, Canada, July 2009.
59. Andrew Morin. On-board Wi-Fi: Maximizing Performance to Meet Passenger Expectations. In *The WiFi on Trains Conference - Train Communications Systems*, London, UK, June 2014.
60. http://www.theatlantic.com/technology/archive/2014/06/you-might-soon-be-able-to-watch-netflix-on-amtrak/372435/.
61. Cosimo Malesci. Fluidity Train to Ground Technology - on-board Wi-Fi at 100 Mbit/sec. In *The WiFi on Trains Conference - Train Communications Systems*, London, UK, June 2014.
62. Hassan Ghannoum, David Sanz, Bernadette Villeforceix, Henri Philippe, and Pascal Mercier. Delivering broadband Internet access for high speed trains passengers using the new WiFi standard 802.11 for train-to-ground communications. In *9th World Congress on Railway Research*, Lille, France, May 2011.
63. Charles D Gavrilovich Jr. Broadband communication on the highways of tomorrow. *Communications Magazine, IEEE*, 39(4):146–154, 2001.
64. Kun-De Lin and Jin-Fu Chang. Communications and entertainment onboard a high-speed public transport system. *Wireless Communications, IEEE*, 9(1):84–89, 2002.

65. Kentaro Ishizu, Masahiro Kuroda, and Hiroshi Harada. Bullet-train Network Architecture for Broadband and Real-time Access. In *IEEE Symposium on Computers and Communications*, pages 241–248, Aveiro, Portugal, July 2007. IEEE.
66. T. Manabe and H. Hojo. Technologies for railway internet services in japan. In *ITST*, Kyoto, Japan, November 2010.
67. Masanori Terada and Fumio Teraoka. Providing a high-speed train with a broadband and fault tolerant IPv4/6 NEMO environment. In *IEEE Globecom Workshops*, pages 1052–1056, Anaheim, US, December 2012.
68. Gaurav Soni and Sandeep Kaushal. Mobile WiMAX technology and its comparison with various competing technologies. *Int. J. Comput. Corporate Res*, 1:1–24, 2011.
69. Marina Aguado, Oscar Onandi, Purificacion Agustin, Marivi Higuero, and Eduardo Taquet. WiMAX on rails. *IEEE Vehicular Technology Magazine*, 3(3):47–56, 2008.
70. Iftekhar Ahmad and Habibi Daryoush. A Novel Mobile WiMAX solution for Higher Throughput. In *16th IEEE International Conference on Networks (ICON)*, pages 1–5, New Dehli, India, December 2008.
71. Sayan Kumar Ray, Krzysztof Pawlikowski, and Harsha Sirisena. Handover in mobile WiMAX networks: The state of art and research issues. *Communications Surveys & Tutorials, IEEE*, 12(3):376–399, 2010.
72. Takayuki Matsumoto. Adding WiFi and Other Information Services to JR East Trains. In *The WiFi on Trains Conference - Train Communications Systems*, London, UK, June 2014.
73. Terry Tse. Study of High-Speed Wireless Data Transmissions for Railroad Operation. Technical report, U.S. Department of Transportation, April 2007.
74. Ting Zhou, Hamid Sharif, Michael Hempel, Puttipong Mahasukhon, and Song Ci. Performance of ieee 802.11 b in mobile railroad environments. In *Vehicular Technology Conference, 2005. VTC-2005-Fall. 2005 IEEE 62nd*, volume 4, pages 2527–2531. IEEE, 2005.
75. Puttipong Mahasukhon, Michael Hempel, and Hsiao-Hwa Cheng. BER Analysis for 802.11b Networks under Mobility. In *International Conference on Communications (ICC)*, pages 4722–4727, June 2007.
76. Jrg Ott and Dirk Kutscher. The "Drive-thru" Architecture: WLAN-based Internet Access on the Road. In *59th Vehicular Technology Conference (VTC-Spring)*, volume 5, pages 2615–2622, May 2004.
77. Richard Gass, James Scott, and Christophe Diot. Measurements of In-Motion 802.11 Networking. In *7th Workshop on Mobile Computing Science & Applications (WMCSA)*, pages 69–74, August 2006.
78. Michal Kowal, Slawomir Kubal, Piotr Piotrowski, and Ryszard Zielinski. Operational characteristic of wireless WiMAX and IEEE 802.11 x systems in underground mine environments. *International Journal of Electronics and Telecommunications*, 56(1):81, 2010.
79. Kazuhiro Yamada. A High Speed Mobile Communication System Implementing Bicasting Architecture on the IP Layer. Technical report, 2012.
80. Bart Lannoo, Didier Colle, Mario Pickavet, and Piet Demeester. Radio-over-fiber-based solution to provide broadband internet access to train passengers. *IEEE Communications Magazine*, 45(2):56–62, 2007.
81. Manfred Schienbein and Juergen Dangelmeyr. TrainCom radio system. In *Telefunken RACOMS*, 2009.
82. http://www.fluidmesh.com.
83. http://www.luceor.com/.
84. http://www.reicom.it/index.html.
85. Seung Nam Choi, Junhyeong Kim, Il Gyu Kim, and Dae Jin Kim. Development of millimeter-wave communication modem for mobile wireless backhaul in mobile hotspot network. *IEIE Transactions on Smart Processing & Computing*, 3(4):212–220, 2014.
86. Tetsuya Arita and Fumio Teraoka. A Fast Handover Mechanism Using Cross-Layer Collaboration for Mobile Networks in High-Speed Trains. In *The third AsiaFI Winter School (Asia Future Internet Forum)*, pages 1–6, Seoul, Korea, February 2010.

87. Shinichiro Haruyama, Hideki Urabe, Tomohiro Shogenji, Shoichi Ishikawa, Masato Hiruta, Fumio Teraoka, Tetsuya Arita, Hiroshi Matsubara, and Shingo Nakagawa. New Ground-to-Train High-Speed Free-Space Optical Communication System with Fast Handover Mechanism. In *Optical Fiber Communication Conference*, Washington, D.C., March 2011. OSA.
88. Masato Hiruta, Masao Nakagawa, Shinichiro Haruyama, and Shoichi Ishikawa. A study on optical wireless. In *11th International Conference on Advanced Communication Technology*, November 2009.
89. Hideaki Kotake, Shinichiro Haruyama, Masao Nakagawa, and Kiyotaka Seki. BER Characteristic of Ground-to-Train Communication System Using Free-Space Optics Technology. In *9th International Conference on Transparent Optical Networks*, volume 3, pages 165–169. IEEE, July 2007.
90. Shingo Nakagawa, Hiroshi Matsubara, Kazuki Nakamura, Daisuke Tatsui, Shinichiro Haruyama, and Fumio Teraoka. Broadband Telecommunication System for Railways Using Laser Technology. In *9th World Congress on Railway Research*, Lille, France, May 2011.
91. Rupak Paudel, Hoa Le-Minh, Zabih Ghassemlooy, and Sujan Rajbhandari. High speed short range optical wireless ground-to-train communications. In *11th Annual Postgraduate Symposium on the Convergence of Telecommunications, Networking and Broadcasting (PGNet)*, Liverpool, June 2010.
92. Rupak Paudel, Hoa Le-Minh, Zabih Ghassemlooy, M Iaz, and Sujan Rajbhandari. High speed train communications systems using free space optics. In *IET Railway Young Professionals Best Paper Competition*, London, UK, October 2010. The Institution of Engineering and Technology.
93. Rupak Paudel, Zabih Ghassemlooy, Hoa Le-Minh, Sujan Rajbhandari, and Bryan Livingstone. Investigation of turbulence effect on the free space optical link for ground-to-train communications. In *11th International Conference on Telecommunications (ConTEL)*, pages 227–232, Graz, Austria, June 2011.
94. Rupak Paudel, Zabih Ghassemlooy, Hoa Le Minh, Sujan Rajbhandari, and Erich Leitgeb. Lambertian source modelling of free space optical ground-to-train communications. In *8th International Symposium on Communication Systems, Networks & Digital Signal Processing (CSNDSP)*, pages 1–5, Poznan, Poland, July 2012. IEEE.
95. Rupak Paudel, Zabih Ghassemlooy, Hoa Le-Minh, and Sujan Rajbhandari. Modelling of free space optical link for ground-to-train communications using a gaussian source. *Optoelectronics, IET*, 7(1):1–8, 2013.
96. A. Nunez, J. Hendriks, Zili Li, B. De Schutter, and R. Dollevoet. Facilitating maintenance decisions on the Dutch railways using big data: The ABA case study. In *IEEE International Conference on Big Data*, pages 48–53, Whashigton, DC, October 2014.
97. R Keith Mobley. *An introduction to predictive maintenance*. Butterworth-Heinemann, 2002.
98. G. Budai, D. Huisman, and R. Dekker. Scheduling preventive railway maintenance activities. In *IEEE International Conference on Systems, Man and Cybernetics*, volume 5, pages 4171–4176, October 2004.
99. Sarah S. Soh, Nor. H. M. Radzi, and H. Haron. Review on Scheduling Techniques of Preventive Maintenance Activities of Railway. In *International Conference on Computational Intelligence, Modelling and Simulation*, pages 310–315, Kuantan, September 2012.
100. D. Prescott and J. Andrews. Modelling maintenance in railway infrastructure management. In *Reliability and Maintainability Symposium*, pages 1–6, Orlando, Florida, January 2013.
101. D. Li. A new maintenance model for high-speed railway signaling system. In *IEEE International Conference on Intelligent Rail Transportation*, pages 239–244, Beijing, September 2013.
102. J.J.V. Antony and G.M. Nasira. Towards predictive maintenance and management in rail sector: A clustering approach. In *International Conference on Recent Trends in Information Technology (ICRTIT)*, pages 502–507, Chennai, July 2013.
103. Y. Han, O.and Same A. Franois, L. Bouillaut, L. Oukhellou, P. Aknin, and G. Branger. Online predictive diagnosis of electrical train door systems. In *World Congress on Railway Research*, Australia, November 2013.

104. Francescomaria Marino, Arcangelo Distante, Pier Luigi Mazzeo, and Ettore Stella. A real-time visual inspection system for railway maintenance: automatic hexagonal-headed bolts detection. *Systems, Man, and Cybernetics, Part C: Applications and Reviews, IEEE Transactions on*, 37(3):418–428, 2007.
105. Esther Resendiz, John M Hart, and Narendra Ahuja. Automated visual inspection of railroad tracks. *Intelligent Transportation Systems, IEEE Transactions on*, 14(2):751–760, 2013.
106. Haoyi Shi, Kaimin Wu, Hengliang Tang, and Shiya Wen. A novel video monitoring system in high-speed railway. In *6th International ICST Conference on Communications and Networking in China (CHINACOM)*, pages 1176–1180, Harbin, August 2011.
107. Shit2Rail. Shift2rail founding members on the Shit2Rail Master Plan. Technical report, Shitf2Rail, 2014.
108. S. Sals. Cloud computing in the service of railway maintenance. In *IET Conference on Railway Condition Monitoring and Non-Destructive Testing*, pages 1–4, Derby, November 2011.
109. Chandra S Bontu, Shalini Periyalwar, and Mark Pecen. Wireless wide-area networks for internet of things: An air interface protocol for IoT and a simultaneous access channel for uplink IoT communication. *Vehicular Technology Magazine, IEEE*, 9(1):54–63, 2014.
110. Ala Al-Fuqaha, Mohsen Guizani, Mehdi Mohammadi, Mohammed Aledhari, and Moussa Ayyash. Internet of things: A survey on enabling technologies, protocols, and applications. *Communications Surveys & Tutorials, IEEE*, 17(4):2347–2376, 2015.
111. P. Rawat, K. Dheep Singh, H. Chaouchi, and J. M. Bonnin. *Wireless sensor networks: a survey on recent developments and potential synergies*. Springer Science, Ovtober 2013.
112. Victoria J Hodge, Simon O'Keefe, Michael Weeks, and Anthony Moulds. Wireless sensor networks for condition monitoring in the railway industry: A survey. *Intelligent Transportation Systems, IEEE Transactions on*, 16(3):1088–1106, 2015.
113. Rong Chen, Ping Wang, and Hao Xu. Integrated monitoring system for rail damage in high speed railway turnout. In *Fourth International Conference on Digital Manufacturing and Automation (ICDMA)*, pages 704–708, Qingdao, June 2013.
114. Shift2Rail. Shift2Rail Scoping Paper. Technical report, Shift2Rail, 2015.
115. S. Ambellouis and J.-L. Bruyelle. *Intelligent Video Surveillance Systems*, chapter Focus on Railway Transport. John Wiley & Sons, 2012.
116. Houssam Salmane, Louahdi Khoudour, and Yassine Ruichek. A video-analysis-based railway–road safety system for detecting hazard situations at level crossings. *Intelligent Transportation Systems, IEEE Transactions on*, 16(2):596–609, 2015.
117. Marco Govoni, Francesco Guidi, Enrico M Vitucci, Vittorio Degli Esposti, Giovanni Tartarini, and Davide Dardari. Ultra-wide bandwidth systems for the surveillance of railway crossing areas. *Communications Magazine, IEEE*, 53(10):117–123, 2015.
118. http://celtic-boss.mik.bme.hu/.
119. http://www.in2rail.eu/.
120. B. Minary and J.-P. Lozach. Train-MD: the Intelligent Transport for Hazardous Goods. In *WCRR*, Montreal, Canada, June 2006.
121. Mathias Grudén, Malkolm Hinnemo, Dragos Dancila, Filip Zherdev, Nils Edvinsson, Kjell Brunberg, Lars Andersson, Roger Bystro m, and Anders Rydberg. Field operational testing for safety improvement of freight trains using wireless monitoring by sensor network. *Wireless Sensor Systems, IET*, 4(2):54–60, 2014.

Chapter 3
Challenges and Perspectives for the Future Broadband Wireless Communications for Railway

This chapter presents the challenges and perspectives for the future broadband wireless communications for railway. First of all, a section is dedicated to the next generation broadband technologies, such as Cognitive Radios and 5G technologies. Then, the current work groups in railway community are presented. Finally, the challenges and perspectives are detailed from different work programs such as in the context of the Joint Undertaking Shift2Rail.

3.1 Next Generation Broadband Technologies

The book aims to describe all the railway applications requiring broadband capacities. We focus on this part on the emerging promising technologies that can solve the problem of increasing needs in terms of throughputs and the problem of spectral congestion.

3.1.1 Spectral Aggregation and Cognitive Radio

3.1.1.1 Definitions and Standards

Among all the solutions for an Internet on board trains, the need of high capacity leads to solutions based on the aggregation of several available frequency bands. This technique is currently used in order to increase the capacity of the systems and allow broadband communications. The satellite and cellular solutions, presented in Chap. 2, are especially based on this aggregation technique. Another solution to optimize the use of the frequency spectrum is to rely on Cognitive Radios (CR).

Recently, the rapid growth in wireless communications has contributed to a huge demand on the deployment of new wireless services. The radio electromagnetic

É. Masson and M. Berbineau, *Broadband Wireless Communications for Railway Applications*, Studies in Systems, Decision and Control 82,
DOI 10.1007/978-3-319-47202-7_3

spectrum is a limited physical quantity, and only a certain part of it is suitable for radio communications. The traditional way of governing this resource has been to administer licenses for portions of the spectrum, usually by a national agency such as US Federal Communications Commission (FCC). Almost all the usable portions of the spectrum are allocated for licensed users. Available electromagnetic spectrum for wireless transmission has become a highly valuable resource. However, recent researches published by the FCC [1] show that the traditional static frequency allocation policy is not efficient and results in poor spectrum utilization. In [2], a general survey of radio frequency bands (from 30 MHz to 3 GHz) is provided.

Figure 3.1 presents the average spectrum occupancy and highlights how low could be the spectrum occupancy in many bands. The dramatic increase in the demand for radio spectrum and the actual low spectral efficiency has spurred the development of a next generation wireless technology referred to as Cognitive Radio. Cognitive Radio concept, introduced by Mitola in 1999 [3], appears to be a tempting solution to the spectral congestion problem by using frequency bands not heavily occupied by licensed users. The International Telecommunication Union (ITU) validated the following definition: "Cognitive Radio system is a radio system employing technology that allows the system to obtain knowledge of its operational and geographical environment, established policies and its internal state; to dynamically and autonomously adjust its operational parameters and protocols according to its obtained knowledge in order to achieve predefined objectives; and learn from the results obtained" [4].

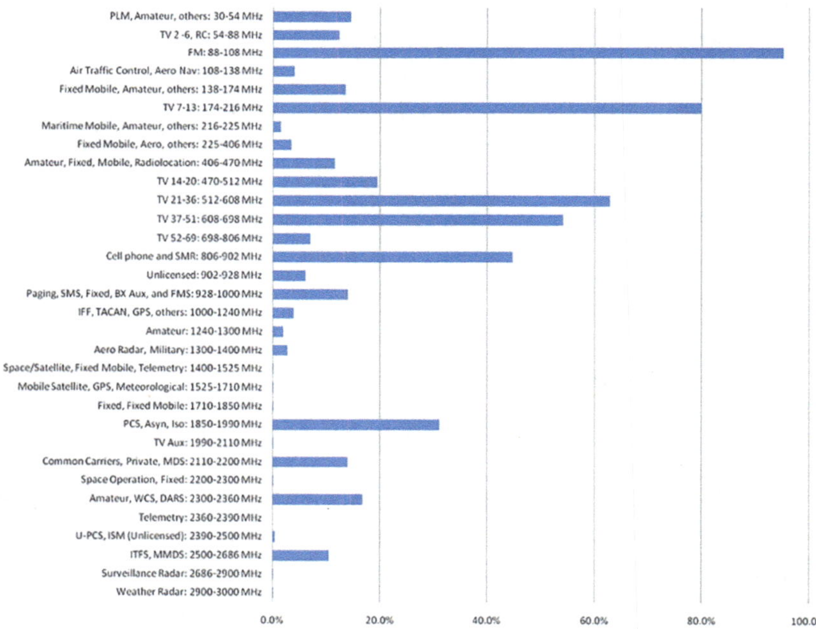

Fig. 3.1 Summary of spectrum band occupancy calculations from [2]

Mitola introduced the cognitive cycle and theorized some concepts which can be summarized as [4]: broad sense adaptation to the environment, intelligence in the network and the terminal, independence of the terminal with respect to the network and the operator and independence of the user with respect to the technique.

A CR system can then adapt its behavior through three steps:

- Information capturing: provided by sensors at all levels to obtain adequate information about radio interface, propagation, network, protocols, security and user requirements;
- Decision making: decisions can be made based on training and/or knowledge bases. The stage of decision making can use information from sensors but also very broad concepts, such as technico-economic considerations via regulatory rules for spectrum use;
- Adaptation: this part concerns the auto-reconfiguration step which is provided by support technology, the Software Defined Radio technology.

Cognitive Radio systems need the development of Intelligent Mobile Terminal, Intelligent Infrastructure and Cognitive Manager that will allow the cooperation between infrastructure and mobile terminal.

Intelligent Mobile Terminal consists of essential stages, such as spectrum sensing, channel estimation, blind modulation recognition or source separation.

Spectrum sensing represents a primary and essential function of CR for dynamic spectrum access. The objective is to detect the white spaces and free frequency bands without causing interferences with Primary User (PU). The spectrum usage can be categorized into black, grey and white spaces. Spectrum sensing relies then on white spaces detection [5]. In [6], a survey of spectrum sensing techniques and algorithms is performed. The algorithms can be classified into three main methods: coherence detection, non-coherent (blind) detection and feature detection. Channel estimation and equalization are also part of the main techniques to set up a Cognitive Radio system.

The Intelligent Infrastructure relies on an inter-layer architecture with potential alternatives for interacting with the heterogeneous access technologies. A basic system of hierarchical data enables learning by memorization in a local database and consolidation in a central knowledge base.

Finally, the decision engine (or “Cognitive Manager”) uses this information to control all components at all levels of the stack (software, operating system, hardware).

In [7], the subject of security issues in the special case of Cognitive Radios is addressed. It is stated that this subject has not yet been many studied compared to other issues for CR case. All communication systems need to be made secure to operate by techniques, such as authentication, authorization, encryption, accounting and non-repudiation processes. Cognitive Radios cause unique security issues, such as the observation of a huge amount of information or the extensive use of collaboration as for the spectrum sensing techniques and the spectrum sharing.

The standardization issues have significant importance in the development of CR systems, since it encourages companies to invest in this domain. Several standards are already published or in a draft status. The most rising one is the IEEE 802.22 standard. IEEE 802.22 is a standard for Wireless Regional Area Network (WRAN) using white spaces in the television (TVWS) frequency spectrum [8].

The development of the IEEE 802.22 WRAN standard aims then to use Cognitive Radio techniques to allow sharing of geographically unused spectrum allocated to the television broadcast service, on a non-interfering basis. It is the first worldwide effort to define a standardized air interface based on CR techniques. The 802.22 working group deals with issues on physical layer and Medium Access Control (MAC) layer and also on spectrum sensing, geolocation and security issues [7]. The novelty of the standard is that it treats at the same time coexistence mechanism between licensed TV band and unlicensed broadband networks, Cognitive Radio concept, super frame structure specially designed for incumbent protection and two layer security concept [9].

The IEEE P1900 standard committee was established as a result of the growing interest for dynamic spectrum access networks. The objective of the committee was to support the standards dealing with dynamic spectrum management and next generation radio developments. The Standards Coordinating committee 41 (SCC41) replace the P1900 since 2007. The seven working groups of the SCC41 are [10]:

- IEEE P1900.1: Definitions and Concepts for Dynamic Spectrum Access: Terminology Relating to Emerging Wireless Networks, System Functionality, and Spectrum Management;
- IEEE P1900.2: Recommended Practice for the Analysis of In-Band and Adjacent Band Interference and Coexistence Between Radio Systems;
- IEEE P1900.3: Recommended Practice for Conformance Evaluation of Software Defined Radio (SDR) Software Modules;
- IEEE P1900.4: Architectural Building Blocks Enabling Network-Device Distributed Decision Making for Optimized Radio Resource Usage in Heterogeneous Wireless Access Networks;
- IEEE P1900.5: Policy Language and Architectures for Managing Cognitive Radio for Dynamic Spectrum Access Applications;
- IEEE P1900.6: Spectrum Sensing Interfaces and Data Structures for Dynamic Spectrum Access and other Advanced Radio Communication Systems;
- IEEE P1900.7: White Space Radio Working Group.

3.1.1.2 Working Groups on Cognitive Radio

Two Cooperation in Science and Technology (COST) working groups deal with Cognitive Radio: the action COST IC0905 TERRA and the action COST IC0902. The first one [11] was active from May 2010 until May 2014. It was organized as a think-tank and aimed to propose regulatory policies and spectrum management solutions that would be conducive to the development of CR and SDR technologies. At the end of these works, a book was published titled "CR Policy and Regulation:

Techno-Economic Studies to Facilitate Dynamic Spectrum Access". Each chapter is dedicated to an action's working group:

- WG1—deployment scenarios: the objective was to collect, analyze and catalog use cases and deployment scenarios relevant to Cognitive Radio;
- WG2—Coexistence (technical rules): the main areas of the WG2 were to explore the role of innovative techniques to facilitate spectrum sharing (situation where two or more radio systems use the same frequency band) and the co-existence (situation where two or more systems operate in adjacent frequency bands);
- WG3—Economic aspects: the WG3 focused on the evaluation of the economic aspects of the developed CR/SDR regulations and on the identification of critical factors that would have significant impact on economic benefits and viability of the proposed regulatory regime for CR/SDR;
- WG4—Impact assessment: WG4 focused on impact assessment for identified combinations of techno-economic sets of CR/SDR deployment rules, with the aim of identifying the most attractive combinations.

Several achievement can be highlighted from the works of the different working groups, such as developed taxonomy of known and future CR/SDR use scenarios and business cases, works on multi-band aggregation, energy efficiency, novel concepts for radio spectrum access, such as pluralistic licensing concept for licensed bands and ISM-Advanced concept (shaping the future of Wi-Fi).

The COST IC0902 [12] started by the end of 2009, dealing with Cognitive Radio and Networking for Cooperative Coexistence of Heterogeneous Wireless Networks. The aim of the COST action is to integrate the cognitive concept across all layers of communication systems, in order to define a European platform for Cognitive Radio and networks. The cognitive concept allows maximizing efficiency in resource management by the coexistence between different heterogeneous wireless networks.

So far, some studies worked on the introduction of cognitive concept at different layers. The COST action aims to go beyond by integrating the cognitive mechanisms across all layers of the system, through techniques such as spectrum sensing, resource allocation and selection between multiple networks. To reach these objectives, algorithms and protocols for all layers of the communications stack aimed to be designed, and a set of standard interfaces and a common reference language for interaction between cognitive network nodes aimed to be defined.

3.1.1.3 Studies and Projects

To our knowledge, only one paper deals with the Cognitive Radio in the Railway context [13]. Cognitive Radios are defined as a cutting-edge research area that combines Artificial Intelligence (AI) with Software-Defined Radios (SDRs), with the goal of improving upon existing radio performance. SDRs are flexible radios because of the realization of some functionality in software. Thus, a Cognitive Engine (CE) uses software-based decision-making and learning algorithms to determine if a change

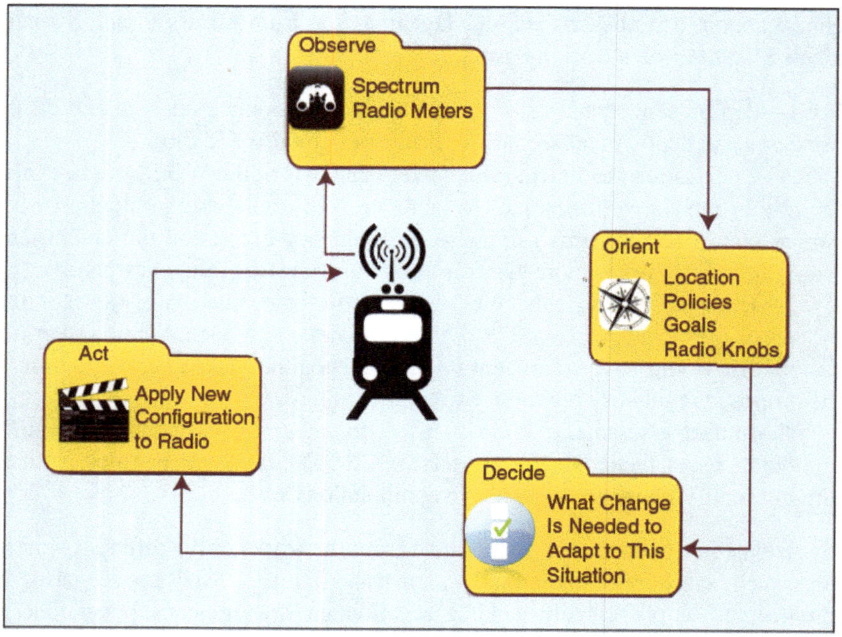

Fig. 3.2 Cognition loop introduced in [13]

in the radio parameters is required. The Cognition Loop is introduced (Fig. 3.2) to explain the Cognitive Radio in the railway context. The decision cycle incorporates then four steps: situational awareness, orientation, decision and action.

Some studies deal with the Cognitive Radio technology. The ICOM architecture, already cited in Chap. 1 was developed in the frame of the INTEGRAIL project, allowing integration of several wireless communication technologies middleware solutions in the railway domain. While the system represents a first step toward system integration and interoperability, the juxtaposition of communication devices along the lines and on board the trains has very high costs. Several European projects can also be cited.

The E3 (end-to-end Efficiency cognitive wireless networks technologies, 2008–2009) project aims to transform current wireless system infrastructures into an integrated, scalable and efficiently managed beyond 3rd generation cognitive system framework. The main issue is to introduce the cognitive systems in the wireless world, while contributing to the standardization of the IEEE P1900.4.

The PHYDYAS (Physical Layer for Dynamic Spectrum Access and Cognitive Radio, 2008–2011) [14] project aimed to propose a dedicated PHY layer suitable for dynamic spectrum access management and Cognitive Radio. Multicarrier techniques are suitable for high throughputs and flexible spectrum allocation. The classical multicarrier scheme is the OFDM, which is largely used in many standards. However, this technique lacks of flexibility and has poor spectral resolution. The project tested the Filter Band-Based MultiCarrier (FBMC) technique which has high spectrum resolution and can provide independent sub-channel with higher throughputs.

The CogEU (Cognitive Radio systems for efficient sharing of TV white spaces in European context, 2010–2012) project aims to build the transition to digital television by developing Cognitive Radio systems which exploit the favorable propagation characteristics of TVWS through introduction and promotion of spectrum trading in real time and creating new frequencies in the upper band of the released spectrum.

The objective of the QoSMOS (Quality of Service and MObility driven cognitive radio Systems, 2010–2013) project [15] was to develop a framework for Cognitive Radio systems and to develop and prove technologies using a test-bed by providing a platform for efficient radio access for future networks.

The EMPhAtiC (Enhanced Multicarrier Techniques for Professional Ad-Hoc and Cell-Based Communications, 2012–2015) project [16] was dealing with Cognitive Radio in the specific case of PMR. The objective of the project was to develop, evaluate and demonstrate the capability of enhanced multicarrier techniques for a better use of existing radio frequency bands and then provide broadband capabilities in coexistence with narrowband legacy services. The project worked also with the FBMC techniques.

Finally, the CORRIDOR (COgnitive Radio for RaIlway through Dynamic and Opportunistic spectrum Reuse, 2011–2015) project [17] aimed to design, develop and evaluate the bricks of a Cognitive Radio system adapted to High Speed Railway to solve the problem of interoperability and costly deployment of incompatible wireless systems along the railway lines and to contribute to spectrum efficiency and global cost reduction. The project works were divided in two main subjects: the intelligent mobile terminal and the intelligent telecommunication infrastructure. For the intelligent mobile terminal, algorithms of spectrum sensing [18], channel estimation [19] and modulation detection [20] were developed. For the intelligent telecommunication infrastructure, the implementation of Cognitive Radio to improve access to spectrum imposes additional features in the protocol stack. Thus, algorithms of spectrum and mobility management and QoS policy were set up with the development of a cross-layer "Cognitive Manager" [21, 22].

3.1.2 5G Next Generation

3.1.2.1 Overview

5G (5th generation mobile networks or 5th generation wireless systems) corresponds to the next phase of mobile telecommunications standards after the current 4G standards. 5G is also referred to as beyond 2020 mobile communications technologies. 5G aims to provide throughputs up to several Gbps. 5G is not yet defined and official, it aims to emerge in 2020. Every big Telecommunication Companies take an interest in this new technology, in addition to the big institutions and the states. The European Union for example financed some programs on 5G such as 5Gnow, IJoin, Tropic and

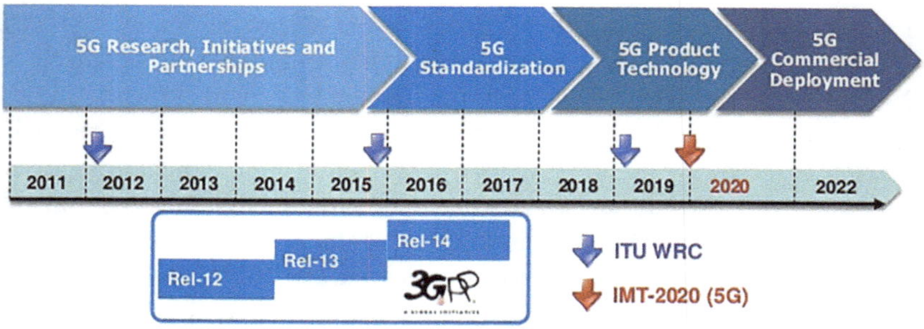

Fig. 3.3 Timeline of 5G technology

the most important: METIS (Mobile and wireless communications Enablers for the Twenty-twenty Information Society) created in November 2012. In China, the IMT-2020 (5G) Promotion group was created in February 2013, in Korea the 5GForum in June 2013 and in Japan, the 2020 and Beyond AdHoc group in October 2013. Furthermore, the Huawei Company announced it had invested in 5G technologies since 2009; South Korea announced an investment of more than one billion euros.

We already cited one typical 5G standard: the WiGig IEEE 802.11ad. Another standard is the LTE-B, the next generation of Long Term Evolution Advanced. It will be developed from 2014 to 2016. It will comprise: LTE Hotspot improvement and small cells, multi-stream aggregation, 3D beamforming and multi-RAT (Radio Access Technology) operation enhancement [23]. Theoretical throughputs of 50 Gbps are expected. The Timeline for 5G is shown in Fig. 3.3.

5G aims to ensure continuity in challenging situations, such as high mobility (e.g. HST) and dense and sparsely populated areas. 5G will also enable to develop Internet of Things by providing a platform able to connect a large number of sensors and devices simultaneously with stringent energy and transmission constraints. Mission critical applications requiring high reliability, such as public safety and railways, will be a part of 5G infrastructure [24]. One unified infrastructure will be able to integrate networking, computing and storage resources, which will also allow to preserve security and privacy. The 5G technology aims to improve performance by increasing capacity, mobility, terminal location, reliability and availability while decreasing latency.

It is important to notice that 5G can be seen as an addendum to the 4G. In other words, the investments done by Railways in 4G will be re-usable when 5G is available. The 5G would bring additional access technologies like millimeter waves mostly dedicated for IoT but should not jeopardize the already deployed 4G infrastructure. Some people or some industry already challenge the fact to adopt 4G for Railways claiming that 5G is almost here.

Two main documents are expected by the ITU to define the 5G [25]:

- IMT.VISION (deadline July 2015): "Framework and overall objectives of the future development of IMT for 2020 and beyond", which objective is to define the

framework and overall objectives of IMT (International Mobile Telecommunications) for 2020 and beyond to drive the future developments for IMT;
- IMT.FUTURE TECHNOLOGY TRENDS (deadline October 2014), which aims to provide a view of future IMT technology aspects 2015–2020 and beyond and to provide information on trends of future IMT technology aspects.

The METIS objectives are to lay the foundation, to ensure a global forum and to build an early global consensus for 5G mobile and wireless communications. The planning is as follows [25]:

- 2012–2014: exploratory research;
- 2015–2017: pre-standardization activities;
- 2018–2019: standardization activities;
- From 2020: commercialization.

In terms of technical challenges, the objectives are:

- A gain of 1000 on mobile data volume;
- A gain of 10 to 100 on the number of connected devices;
- A gain of 10 to 100 on the throughputs (up to 10 Gbps);
- A diminution of 5 on the end-to-end latency;
- A gain of 10 on energy consumption with higher battery life relying on low-power devices.

The Fig. 3.4 presents the challenges and scenarios of the 5G, which summarizes the objectives of the emerging standard concerning throughputs, number of users, mobility, latency and low-power devices.

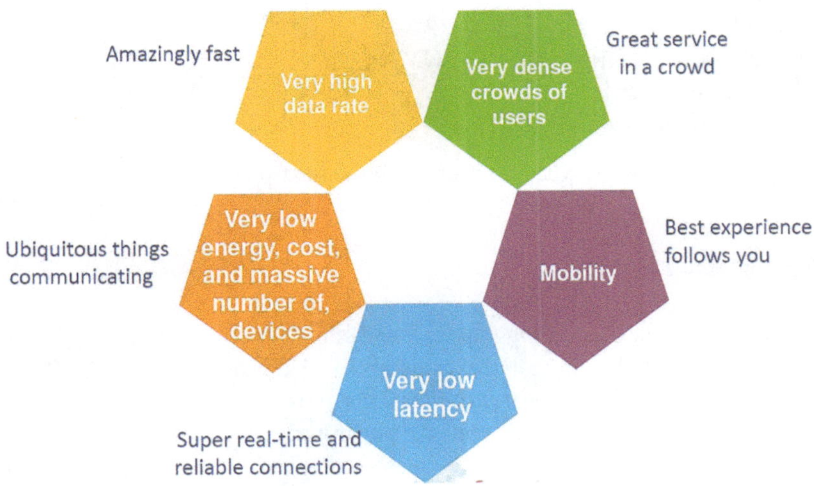

Fig. 3.4 Challenges and scenarios of the 5G from [25]

3.1.2.2 Key Technological Supports of 5G

The 5G wireless technology will support a set of heterogeneous wireless technologies, from evolution of current access schemes to development of new technologies, relying on cellular and satellite solutions. 5G will rely on seamless handover mechanisms between the different access technologies and simultaneous radio access technologies in order to increase reliability and availability.

5G will then require the deployment of Ultra-Dense Networks (UDN) with several small cells. Techniques of interference mitigation and backhauling are considered and 5G will be driven by software. Some emerging technologies are considered to reach the objectives of performance, scalability and agility, such as Software-Defined Networking (SDN), Network Functions Virtualization (NFV), Mobile Edge Computing (MEC) and Fog Computing (FC) but also Data Analytics and Big Data.

Regarding spectrum management, very wide contiguous frequency bandwidths will be required to fulfill the high capacity needs. To achieve these bandwidth requirements, higher frequencies above 6 GHz are considered. These new bands will be carefully validated by especially studying the other services that use or plan to use the bands. It will be also essential to take into account long-term investments so that they can be preserved.

The key technologies and principles in 5G wireless transmission are [26]:

- Use of contiguous and wide spectrum bandwidth;
- Flexible resource allocation and sharing schemes;
- Flexible air interfaces and new waveforms;
- Agile access techniques;
- Advanced multi-antenna MIMO techniques, such as massive MIMO, beamforming and beamtracking;
- Full-duplex;
- Non Orthogonal Multiple Access (NOMA);
- Enhanced Multi-carrier;
- Advanced coding and modulation.

A zoom on massive MIMO is presented in the next section.

3.1.2.3 Massive MIMO

MIMO technology has becoming mature for wireless communications and has been incorporated into wireless broadband standards like LTE and Wi-Fi (IEEE 802.11n). Basically, the more antennas the transmitter/receiver is equipped with, the more there are possible signal paths and the better the performance are in terms of throughput and link reliability. The price to pay is increased complexity of the hardware (number of RF amplifier front-ends) and the complexity and energy consumption of the signal processing at both ends.

Massive MIMO (also known as Large-Scale Antenna Systems, Very Large MIMO, Hyper MIMO, and Full-Dimension MIMO) consists in the use of a very large number of antennas (e.g. hundreds or thousands) that are operated fully coherently and adaptively. It offers large network capacities in multi-user scenarios. The large number of antennas allows focusing the transmission or the reception into small regions of space. It leads to huge improvements in terms of throughput and energy efficiency. Massive MIMO was originally planned for Time Division Duplex (TDD) operations, but can also be applied in Frequency Division Duplex (FDD) ones [27]. The main benefits of massive MIMO are the extensive use of inexpensive low-power components, the reduced latencies, the simplification of the MAC layer, and the robustness to the interference and intentional jamming [28, 29]. Massive MIMO can increase the capacity 10 times or more, by an aggressive spatial multiplexing, and simultaneously improve the radiated energy efficiency on the order of 100, thanks to a large number of antennas focusing the energy into small regions in space [28].

Massive MIMO is an important research topic, which has been mostly theoretical so far. Nevertheless, first basic tests beds are available [30] and initial channel measurements were implemented [31, 32].

Massive MIMO leads to many classical problems in communication theory but also uncovers entirely new problems, which represent a gold mine of research problems [28]:

- A fast and distributed coherent signal processing, due to the amounts of data to be processed in real time;
- The challenge of low-cost hardware, due to the huge number of used antennas;
- The problem of hardware impairments due to low-cost components leading to high phase noise;
- The channel characterization due to the additional properties of the channel provided by the Massive MIMO;
- The cost of the reciprocity calibration needed with required TDD;
- The pilot contamination, which makes difficult the uplink detection and downlink precoding.

As described, the massive MIMO has a large potential as a key for 5G next generation systems.

3.1.2.4 Timeline of 5G

5G is expected to be commercialized by 2020. Japan has committed to have 5G commercial services for the Olympics in 2020. The time period between 2014 and 2020 is dedicated to research and standardization processes. The 5G timeline is detailed in the following [24]:

- 2014–2015: it represents the exploratory phase. The detailed requirements on 5G systems are studied in order to identify the most promising functional architectures and technological options to meet the requirements. All these exploratory works are carried out in the framework of collaborative projects in industry and research programs;

- 2015–2017: it represents the research and development phase. The detailed studies on all access means, backbone and core networks will be initiated, by taking into account economic conditions for future deployment;
- 2016–2018: it represents the detailed system optimization. The requirements and constraints will be taken into account. The works on the envisaged frequency bands will be initiated and the final system definition and optimization will be performed, relying on simulations, validation of concepts and first trials. All these works will represent initial contributions to global standardization activities;
- 2017–2018: it represents the phase of investigation, prototypes, and technology demonstrations of network management and operation. Simulations and trials of systems concepts will feed the detailed standardization process;
- 2018–2020: it represents the phase of demonstrations, trials and scalability testing depending on standard readiness and component availability;
- 2020: the new frequency bands are expected to be available for trial network deployment and initial commercial deployment of systems.

3.1.3 Satellite Technologies

Chapter 2 presented the Internet on board train solutions relying on satellite technologies. Several technologies relying on satellites are researched and refined through a wide range of initiatives around the world [33]. They are presented in this section. One can note that the perspectives are multi-faceted involving technology, economics, global and regional regulations, politics and societal trends. Only the technical aspects will be presented in this part.

The evolutions of satellite technologies are influenced by parallel evolution of others technologies or sectors, such as terrestrial mobile communication segment. It is then recognized that it is very beneficial to align the initiative in terms of research and development between terrestrial mobile communications and Mobile Satellite Systems (MSSs). Several capacity enhancement can be then considered, such as multiuser detection, advanced frequency planning techniques, cross-layer optimization of radio resources, and Cognitive Radio. The objectives of the research and development are to improve the spectral efficiency in order to maximize the utilization of the spectrum and reduce the cost/bit.

The Multi-User Detection (MUD) was initially studied and used for CDMA scheme. The technique is used to detect the users jointly by minimizing the Multiple Access Interference. Such technique is able to enhance spectral efficiency.

Advanced frequency planning represents a technique to enhance spectral efficiency. Frequency spectrum of Geostationary MSSs is shared among operators and is quite fragmented. It is then crucial to maximize the utilization of the available spectrum and to optimize the spectrum efficiency. We can mention the L and S bands which are in high demand and therefore congested. It was proven that optimized algorithms of frequency planning relying only on a subset of algorithms present many advantages. Works were for instance performed for the Inmarsat system, known to have the most complex and dynamic frequency planning environment [33].

IP enabled solutions lead to an increase of data communications. However, IP is based on a rigid hierarchical protocol architecture without cross-layer interaction. However, for wireless communications, cross-layer techniques are quite efficient to optimize the overall resource utilization.

Finally, Cognitive Radio techniques presented in this chapter are quite suitable for MSSs. Congestion in frequency bands in satellite communication bands (L, S, C, X and Ku bands) leads to encourage methods and techniques to maximize and optimize the use of these bands. An approach relying on Cognitive Radio represents a quite good solution to mitigate the congestion for satellite systems. A study conducted by ESA showed that the concept can be applied to satellite communications in many ways [33]. The possibilities are listed below:

- The satellite downlink signal can not be used in the context of CR due to the wide area covered by the satellite, which prevent the satellite to be used as a secondary user. On the other side, the uplink signal can be used and works on spectrum sensing of satellite system are only performed in the uplink direction;
- The mobility is another important aspect of MSS for CR;
- For hybrid systems, satellite elevation angles add a reusability dimension;
- A CR system from satellite to satellite can be considered, relying on a shared central dynamic spectrum database.

As seen in Chap. 2, the trend for the Internet on board systems relying on satellite technologies is to go to the Ka band. At last, current studies at ERA, that will be seen in the following, highlighted the possibility of using MEO satellites by 2030 for operational applications.

3.1.4 Other Future Technologies

As presented in this section, 5G networks will rely on several optimization techniques and concepts, such as Software-Defined Networking (SDN), Network Functions Virtualization (NFV), Mobile Edge Computing (MEC) and Fog Computing (FC). They are presented in this part.

SDN is an emerging architecture that allows network administrators to manage network services through abstraction of lower-level functionality. The system that makes decisions about where traffic is sent (the control plane) is then decoupled from the underlying systems that forward traffic to the selected destination (the data plane). SDN requires some methods to allow communication between the control plane and the data plane. The architecture is dynamic, manageable, cost-effective, and adaptable, and so well suited for high bandwidth and dynamic nature of today's applications.

NFV (also known as Virtual Network Function (VNF)) is a network architecture concept that offers a new way to design, deploy and manage networking services. The solution relies on IT virtualization related technologies to virtualize entire classes of network node functions into building blocks that may be connected, or chained, to create communication services. A VNF may consist of one or more virtual machines running different softwares and processes, on top of industry standard high volume servers, switches and storage, or even cloud computing infrastructure, instead of having custom hardware appliances for each network function. Thus, NFV decouples the network functions from proprietary hardware appliances so they can run in software. It is applicable to any data plane processing or control plane function in both wired and wireless network infrastructures.

MEC is a system that offers application developers and content providers cloud computing capabilities and an IT service environment at the edge of the mobile network. This environment is characterized by ultra-low latency and high bandwidth as well as real-time access to radio network information that can be leveraged by applications. MEC provides a new ecosystem and value chain. Operators can open their Radio Access Network (RAN) edge to authorized third-parties, allowing them to flexibly and rapidly deploy innovative applications and services towards mobile subscribers, enterprises and vertical segments.

FC is an architecture that uses one or a collaborative multitude of end-user clients or near-user edge devices to carry out a substantial amount of storage (rather than stored primarily in cloud data centers), communication (rather than routed over the Internet backbone), and control, configuration, measurement and management (rather than controlled primarily by network gateways such as those in the LTE).

3.2 Current Works and Discussions

3.2.1 Works on Professional Mobile Radio

Works and discussions are taking place currently to study and evaluate the future for Professional Mobile Radio. While current PMR networks are used to essentially carry low throughput voice and data flows, the question arises for the future developments, particularly in the context of wider development of mobile communications networks towards the provision of broadband data transmission services [34].

Several PMR systems are deployed for different applications. We can distinguish two main categories:

- The digital technologies with band lower than 25 kHz, providing throughput up to few tens of kbps, for example TETRA technologies TETRAPOL, DMR, dPMR, NXDN;
- The digital technologies operating in channels form 50 to 200 kHz, providing throughput up to several hundred kbps, for example TEDS and GSM-R technology.

The evolution of PMR uses to high-speed services raises questions about the technologies that can meet these future new needs. Indeed, the current deployed PMR ensures the voice communication needs and short messages data type. However, the technology does not seem at this stage able to address the problem of high or very high-speed data transfers. It appears that suppliers of "classical" mobile network equipment and PMR specialists actors have entered into partnerships for the development of PMR solutions based on the LTE technology, including communications infrastructure and mobile devices. LTE seems to be one of the possible evolution of technology to meet all or part of PMR needs, with larger bandwidth of 1.4 MHz, 3 MHz or 5 MHz [34].

The AGURRE French organism [35] aims to gather, represent and defend the interests of the major users of PMR in order to define a regulatory framework for the future implementation of broadband networks. AGURRE brings together different stakeholders, such as airline companies, airports, railway or highway operators. The approach consists in the participation in different groups of national and international works that are responsible for establishing a unified technology standard for broadband radio networks. AGURRE ensures that:

- All the professional needs are ensured by the standard (transport, security, industry, ...)
- The standard is recognized at the world level (Europe, North America, Asia, ...)
- The standard is part of a common technology for both the mass market and the professional market.

The objective is also to propose a national translation of the technology in France that is suited to the professional world, by the allocation of frequency spectrum bands and the possible use of these bands by users.

The implementation of broadband radio networks aims to ensure functional needs, such as:

- Video streams transmission in mobility conditions;
- Connection of mobile equipments (vehicles or persons) to a centralized computer application.

The latter application can be used for instance for the transfer of maintenance or for ticketing data. These needs fit naturally into a people and objects connectivity trend that can be generalized to the whole society.

The strategy of AGURRE is to rely on widespread technologies and to offer maximum interoperability at both terminals and network infrastructures. Therefore, the association looks at the LTE technology. Regarding frequency spectrum resource, AGURRE discussions lead to the obtaining of a "principal" frequency band which could be contained in one of these two bands:

- The 400 MHz band, currently used by PMR;
- The 700 MHz band, which will be released soon (as it will be seen in the following).

Other "secondary" bands could be allocated to answer specific requirements, such as transmission in confined areas such as tunnels, which will need high frequency band (>2 GHz), or transmission on long distances on lower frequency bands

(<380 MHz). AGURRE is currently working to evaluate the bandwidth required to ensure functional needs expressed by users.

Finally, it is important to notice that USA Public Safety decided several years ago to adopt LTE 700 MHz as the future of the current TETRA technology. Furthermore, the TETRA MoU promotes the fact that TETRA technology can be maintained as a narrow band technology for critical services and complemented by a LTE providing broadband for complementary non-critical services.

3.2.2 Works in Railway Community

As presented in Chap. 1, ERA, UIC and ETSI are currently working on the evolution of the GSM-R technology, getting feedback from Railways operators and GSM-R Industry Group. Globally, UIC takes care of User Requirements in general, ERA takes care of interoperability considerations at EU level (including spectrum), and ETSI shall be the vehicle for standardisation. The works are presented in this section.

3.2.2.1 Works at ERA

ERA, as System Authority for ERTMS, leads the essential activities to introduce the new radio system that will replace the GSM-R. Three main motivations lead to think about the future of GSM-R:

- Obsolescence of GSM-R is expected for 2030 as there will be no more support for technology. It is then crucial to consider right now the technology that will replace the GSM-R;
- The needs in terms of functionality and bandwidth have changed. According to ERA, these changes are slow, contrary to the views expressed by some stakeholders;
- A cost reduction is also expressed, essentially the costs in terms of OPEX. It would be wise to avoid having a new niche market with LTE-based railway radio products.

ERA is in charge of TSI-CCS (Technical Specifications for Interoperability—Control Command and Signaling). It is the authority that manages the ERTMS. As per the European regulation establishing the Agency, ERA is the system authority for ERTMS. The necessary provisions in the CCS TSI enabling migration of technologies that can be used by the trackside and on-board from GSM-R to a next generation system will be then introduced.

The identification of the necessary changes to the current legal framework is expected by the European Commission in 2018. The replacing system has to be available for deployment in 2022. The main applications of the current radio system, as defined in the CCS TSI, are considered as reference and co-existence with these applications is mandatory. However, the system shall be capable to support future changes and extensions of the applications covered by the CCS TSI and other TSI's, as well as railway related applications outside TSI's.

The overall program plan covers all activities to be performed by ERA that are essential to update the CCS TSI to allow other system(s) in coexistence of GSM-R. The Program consists then of the following steps:

- Performing studies on relevant subjects and define options;
- Select the relevant option(s) and provide justification (including impact assessment);
- Conclude on Requirements and Specifications, to be included in the CCS TSI.

After 3 workshops organized at ERA on the development of rail systems which took place at the end of 2014 and in early 2015, a first report on the evolution of the GSM-R was published in April 2015 [36].

ERA has developed this study to evaluate the possible options for development of railway operational communications in terms of:

- Methodology for assessment;
- Feasibility of the options presented;
- Selection of the most appropriate options;
- Possible concepts for rail operators depending on selected options;
- Availability of frequencies for railway, analysis for the different applications and possible use of a common bandwidth for different services;
- Evolution of terminals and network infrastructure depending on the options;
- Highlighting elements for the economic evaluation of the options studied.

Multi-layered decisions can be set up:

1. A cross-sectoral decision at European level on how to manage comprehensively the Public Protection Disaster Relief (PPDR), energy and railway communications;
2. A specific European decision for rail: these decisions are made at the European Commission (DG MOVE) and at ERA. Other stakeholders may also be involved;
3. Decisions at Member States (national) level: Coverage may depend on the particular type of line in terms of traffic and speed.

All the study relying on multi-layered process, considering the different options regarding functionality and so on lead to a first choice is performed for the future of the GSM-R technology. It relies on independent application bearers that run on a range of enabling technologies, including potentially:

- Mobile technology enabling data such as GPRS and EDGE technology;
- The current and future LTE;
- Possibly existing and future Wi-Fi technology;
- Possibly satellite technologies (for lines where conditions are appropriate).

ERA is particularly interested in the feasibility to use satellite communication technologies for railway applications. ERA is concerned about the feasibility of existing satcom services and products to meet the requirements of railway voice and data applications, on the geographical part of the European railway network where such satellite technologies could be used and also the economic impact compared with terrestrial solutions.

Other technologies may also be considered such as 5G when it will be defined.

Major points to be resolved are the coexistence between the different communication networks, based on GSM-R and other technologies, the continuity of the voice and data services, as well as the strategy to make the architecture chosen future proof (to reduce costs when a new technology will be added).

3.2.2.2 Works at UIC

Future Railways Mobile Communication System (FRMCS) [37] is a UIC project launched in 2014 for three years, after five years of preliminary investigations. FRMCS is the provisional name of the system which will succeed to GSM-R. The objective is to have FRMCS available in 2022.

For the choice of future radio system, the objectives of the FRMCS project is to take into account:

- The user needs;
- The spectrum availability versus needs;
- The operational impact, migration strategy;
- The technology.

FRMCS is divided in 3 Working Groups:

- The Functional WG which already produced a User Requirement Specification for the GSM-R successor;
- The Architecture and Technology WG, which considers all technologies possibilities and defines a system architecture relying on a bearer agnostic approach;
- The Spectrum WG, which estimates the amount of bandwidth required and the spectrum options.

The works pursued in the FMRCS project already provided some first results and conclusions.

Regarding the Functional WG, the user requirement specifications was delivered. Next steps consist in building of the use cases, and preparing the functional requirement specification.

The Architecture and Technology WG already some candidates for the choice of the technology:

- 4G, 5G;
- Possible satellite communications for low traffic rural lines;
- Possible Wi-Fi to increase capacity in stations.

The Spectrum WG drafted a document to initiate work to understand better if and how sharing of spectrum and maybe more with the "blue light" community may work. Given spectrum scarcity and price of the bandwidth, FRMCS project is studying different options, which are not only related to frequency but also to network model and system architecture, with some technology choices.

The works highlight that change is necessary, it has to and it will be done at a certain time. The railway community must use the GSM-R lessons learned. Migration must be smart. Multimode Radio will be the key for migration and for technology flexibility. The new system must be as good as the existing one, and cost effective.

3.2.2.3 Works at ETSI

ETSI Next Generation Radio for Rail (NG2R) is the forum for Industry and Railway Operators to confront ideas on the GSM-R successor. It is also in charge of producing a System Reference Document to liaise with Electronic Communication Committee (ECC)/CEPT on spectrum needs.

The objectives of NG2R are:

- To provide standards for voice, data services and other applications over broadband and narrow band air interfaces for the Rail Transportation domain;
- To collect requirements from relevant stakeholders from the Rail Transportation domain, including urban, suburban, regional, long distance for PMR Access systems;
- To contribute to develop ETSI existing and future standards to allow for taking into account these additional specific requirements.

The activities of NG2R will cover in particular the following areas:

- To address the requirements for mission critical and business related services as defined in particular by the UIC FRMCS WGs;
- To identify and fill in standardization gaps covering:
  - Architectural design for end-to-end mission critical communications services for transport to be delivered over mobile radio systems;
  - Standard development for services and interfaces for transport applications;
  - Appropriate user equipment (on-board, handheld, stationary, special purpose radio, communication gateways, etc.) for critical audio, data and video applications.
- To identify further development of the standards applicable to rail transport allowing radio performance enhancements and resource optimisation in the field of:

  - Improving system capacity;
  - Improving the spectrum efficiency;
  - Increasing the robustness of the radio link.

- To identify any spectrum needs for the Next Generation Radio for Rail System.

Finally, it is important to notice that all these organizations are not only concerned by Signaling but also by operational and passengers services. This gives a complementary view on the requirement and specification (in addition to ETCS and Operational voice targeted by ERA) having an impact on the choice of broadband technology.

3.2.3 *What About Current Works on Internet on Board in France?*

As presented in Chap. 2, SNCF, the French National Railway Operator, performed many tests to provide Internet on board HST [38–40]. Measurements and experiments were performed with Wi-Fi technologies and the "BoxTGV", relying on satellite technology, was also launched on the TGV-Est line, but stopped for profitability reasons.

Currently, SNCF launched a call for tenders in order to:

- Expand the coverage of 4G cellular networks along the railway tracks. This one concerns then the mobile network operators. The main Telecom operators have then been assigned for the increase of 4G coverage along the lines;
- Develop a solution to provide an Internet access on board HST, relying on Wi-Fi deployed in the carriages and connection provided by cellular networks.

The Wi-Fi relying on 4G networks is then planned in HST for 2017. Initially and to our knowledge, requirements in terms of throughput are to reach 15 Mbps and evolve to 25 Mbps. At the same time, the french frequency regulator, ARCEP, fixed its major strategic objectives in terms of 4G coverage for trains for the coming years. Thus, a 4G coverage of 60 % of regional trains is planned by 2020, 80 % by 2027 and 90 % by 2030. This timeline is linked to the allocation procedure of the new 700 MHz band for 4G, which will be spread out from 2016 to 2019.

Discussions and decisions are still on going on the call for tenders presented above. The connection between train and ground could be eventually provided only by a direct connection of passengers via their terminal (smartphone, tablet) to the 3G/4G cellular public networks on the majority of the lines. The French railway operator intends to push MNOs in order to improve the cellular coverage of the railway tracks.

Thereby, French railway operator opted for a solution relying on public cellular networks, like many other railway operators. Indeed, it represents a first reasonably cheap solution to ensure connection on board trains. However, limitations in terms of capacity and throughput are still an issue and other solutions could be developed to allow broadband links and ensure all the possible applications presented in Chaps. 1 and 2.

3.3 Challenges and Perspectives

3.3.1 European Commission Objectives

The European Commission has to address the major societal issues for a more competitive European transport system, such as rising traffic demand, congestion, security of energy supply and climate change [41]. A White Paper of the Commission was then published in 2011 and presented a set of objectives to reinforce railway transportation:

- For rail passenger:
  - By 2020, establishment of the framework for a European multimodal transport information, management and payment system;
  - By 2030, triply of the length of the existing High Speed rail network for, by 2050, a majority of medium-distance passenger transport by rail (outpace aviation for journeys up to 1000 km);
  - By 2050, connection of all core network airports to the rail network, preferably High Speed.
- For freight transportation:
  - By 2030, translation of 30 % of road freight over 300 Km to other modes such as rail or waterborne transport, and more than 50 % by 2050;
  - Doubling of rail freight compared to 2005;
  - By 2030, deployment of ERTMS on the European Core Network;
  - By 2050, connection of all seaports to the rail freight system;
  - Rail Freight Corridors as the backbone of the EU freight transport system.

The overall challenge of the European rail sector is the strengthening of the rail sector in the European transport system, by facing and outpacing road and air sectors. To do this, the sector has to face a number of challenges [41]:

- A Quality of Service and attractiveness challenge, by increasing the Quality of Service of the rail and improving services and customer quality, for both rail passenger and freight;
- A cost and competitiveness challenge, by reduction life-cycle cost of the railway system, and creating for instance new business opportunities;
- A fragmentation of national rail markets challenge, by developing the interoperable unique European rail system;

- An infrastructure challenge, by the need to invest in new solutions to overcome the environmental issues, such as noise and vibration;
- A know-how challenge, with the challenge of the availability of skilled people;
- An innovation challenge, by taking advantage of rail advantages in terms of environment performance, safety and efficiency.

3.3.2 Scientific Barriers

Regarding all the subjects, technologies and challenges presented in this chapter, several scientific barriers can be identified. First of all, the emphasis was placed on the spectrum management, which represents a great challenge regarding the necessity to optimize the utilization of the frequency spectrum.

For instance, dynamic allocation of frequency spectrum is initiated in the CR works presented above. A large number of wireless communication systems are operating at different frequencies and railway system's efficiency would be improved by integrating all these heterogeneous wireless networks.

A second important feature is to take into account the Electromagnetic Compatibility (EMC) aspects. The multiplication of telecommunication systems in the transportation environment results in the increase of potential sources of electromagnetic interference. In addition, the continued diminution of power levels necessary for the operation of electronic systems tends to make them more vulnerable.

The EMC represents then a foundation to improve the availability, reliability and integration of communication devices. EMC can rely on characterization and modeling of the electromagnetic environment and the behavior of electronic equipment and telecommunication systems.

EMC activities deal with interferences that can be intentional or not. In the former case, we talk about cyber-attacks. Systems have then to be equipped with a cybersecurity layer. Due to the amount of sensitive information on networks and the vulnerability and porosity of railway systems, an optimal level of cybersecurity has to be achieved to fight the different threats for the railway signaling and telecommunication systems. Cybersecurity has also to fight cyber-attacks and advanced persistent threats from outside. The challenges are to set up a system that is sustainable, integrated, interconnected and supported by the network.

Cybersecurity systems rely on threat detection, identification, modeling and monitoring. The different threats can be electromagnetic jamming, eavesdropping, denial or service attacks, spoofing attacks and equipment infection attacks. The fight against attacks can rely on the adaptation of security layers of the communication systems, suitable for railway context.

As presented in this section, SDN can be used as an enhanced technology to set up a high performance and scalable intrusion detection system and to block traffic. Works on physical layer and multipaths can also be performed to resist to jamming for instance. In another hand, cybersecurity vulnerability increases when using standards components on the market, but also implementing large systems combining critical and non-critical applications or multiple/combined networks. So enlarging the radio paths capability and the system complexity/configurability might also enlarge the vulnerabilities.

The security in the context of Cognitive Radio was also addressed in this chapter. Indeed, Cognitive Radios cause unique security issues, due to the observation of a huge amount of information or the extensive use of collaboration for the spectrum sharing for instance. Finally, innovative solutions for cryptography and safe key management systems can be developed to reduce complexity of large-scale systems, such as railway systems with a large number of connected objects.

A complete policy has to be defined and set up on the cybersecurity issues for railway domain. The current networks are usually heterogeneous and not protected. The communication systems already standardized are not flexible, cost effective and sustainable with an already planned obsolescence of the used cryptographic techniques. The set up of a security system will ensure high availability, authentication and integrity of the railway systems. Safety and security are quite related and safety can not be ensured without security.

Finally, it is important to notice that the telecom system is not only radio path, but also onboard and wayside networks connected by routers. So combining the radio path has an impact on the complete chain and may provide risk to link a critical domain with a non-critical domain. Moreover, the cybersecurity objectives may give severe constraints to the systems, like some potential physical segregation, or access limitation or configuration limitation. Cybersecurity must then be a key driver for the telecom evolution.

Energy management represent a big challenge for the railway sectors. The trends are to reproduce models from other sectors and adapt them to the railway one. For instance, some works on smart metering are initiated in the context of the In2Rail European project, as presented in Chap. 2. The objective is to manage the energy flows in order to minimize the global consumption.

3.3.3 Key Challenges Extracted from the Shift2Rail Multi-annual Action Plan

Globally, current railway systems do not sufficiently take advantage of new technologies, such as high capacity wireless communication systems, but also satellite positioning and innovative real-time data collection and processing. The vision for the future of the railway, in the context of the Shift2Rail initiative [42], is to develop smart and fail-safe communication systems.

3.3.3.1 Adaptable Communications

The first objective is to specify an IP-based communication system, able to combine heterogeneous wireless technologies, such as LTE, Wi-Fi or satellite. Such systems have to be able to enhance throughput and safety-security features. They also have to be resilient to interferences and radio technology evolution. As presented in this chapter, key future technologies are the Cognitive Radio and the 5G wireless technologies.

The overall concept is then to develop technology-independent adaptable train-to-ground communication systems. Threats such as interference and cyber-attacks have to be taken into account in the design of the communication systems. Several key technologies have to be set up, such as 5G, SDN, NFV, KPI evaluation systems, SDR platforms, IP-based communication gateway able to perform bandwidth aggregation, dynamic spectrum allocation and mobility support (key components of the CR).

Given the current and fast evolutions of mobile wireless communication standards, such as the works regarding GSM-R evolution and the current standardization process of 5G technologies, it is necessary to develop a system able to successfully integrate a number of heterogeneous technologies and communication protocols into a single network in order to take advantage of the various deployments provided by external network operators and dedicated infrastructure. The objective is to shift from the "Network as a service" to "Network as an asset" model vision. CAPEX and OPEX can then be minimized. The migration can be performed by relying on middleware platforms for transparent switching radio components.

3.3.3.2 Virtual Coupling

Many European railway lines need to increase the capacity of their lines, which are very busy. New lines and tracks can be built and added but it is quiet expensive and it will not allow to fulfill the future capacity requirements. The concept of virtual coupling can be then introduced to allow more flexible train operation and better utilization of existing railway infrastructure and will be part of the Shift2Rail work program.

The concept of virtual coupling relies on the exploration and demonstration of Virtually-Coupled Train Sets (VCTS). It allows to operate physical trains much closer to one another (inside the absolute braking distance) and to modify dynamically the composition of the trains. The convoy of virtually coupled trains is then linked by a ultra-reliable hard real-time communications radio, sharing the same data (speed, braking commands, etc.) [43]. The concept is basically to reduce the headway and then to increase the capacity on the lines. It can be seen as the extreme limit of the concept of moving block, removing the "one block, one train" limitation. The functionality relies on board environment in addition to new features in the wayside signalling and supervision systems.

The application of the virtual coupling concept involves several steps, such as the analysis of radio aspects, related to the direct radio communications within the

VCTS, through external network and what radio systems are available and the related functionality.

3.3.3.3 IP-Based WSN Architectures

Another objective is to develop smart wireless communication equipment able to connect wayside objects. The objective is then to demonstrate solution of object controllers realizing a decentralized approach to rail automation. Although the trend for future signalling systems will be to reduce and even remove trackside equipment, this wireless trackside architecture remains relevant, at least for interfaces to level crossings.

Currently, trackside objects are mainly interfaced cables, which is expensive, vulnerable and complex. A solution relying on locally derived power and radio communications with maximum decentralisation can then overcome these drawbacks.

Safe, secure and locally powered IP-based radio communication and low energy consumption WSN are the key components of these systems able to perform collection of data from thousands of fixed sensors.

The IP-based WSN architecture can be based on adhoc, mesh or relaying approaches. The enhancement of wireless technologies for the trackside sensors will able the interoperability between several heterogeneous WSNs already deployed and the collection and transmission of a huge number of data from the sensors.

Such architectures will allow the optimization of global railway operation by facilitating decision-making process and achieving decentralized wireless control of remote trackside objects.

3.3.3.4 Wireless TCMS

The TCMS is the communications backbone of the train, which has some essential roles on vehicle performance, such as the integration and management of on boad information, the communication between equipment, between vehicles and with the ground, and the integration of the different subsystems of the train.

The technologies used for the TCMS are very specific to the railway domain and existing standard for train control functions are quite limited, and the increase in information volumes and new diagnostics and passenger oriented services requires a more efficient network.

Current practice in industry (such as SNCF in France) relies on a standard TCMS network for train control functions and an additional network for non critical other functions. This implies considerable amount of cables and then weight and complexity. Furthermore, integration of subsystems and commissioning of TCMS require huge efforts and take very long time due to the lack of standardised application profiles and appropriate architectures, simulation and testing frameworks.

Current researches are dealing with new generation of train communication systems based on wireless transmission for TCMS especially. The objective is to reduce or remove all on board communication cables and simplify the train coupling procedure. For instance, the objectives of the European projet Roll2Rail is to increase the operational reliability (up to 50 %) and to reduce the life-cycle costs (up to 40 %).

The next generation of wireless TCMS has then the objectives of increasing the reliability, while keeping the current levels of safety and security of wired systems. Specific architectures and protocols for safety related functions and safety data transmission have to be set up.

3.3.4 Summary on Expectations and Challenges of Railway Domain

The railway community faces many challenges in order to perform railway digitalisation. In [44], the Community of European Railway (CER), the association CIT (association of railway undertakings and shipping companies), the association of European Rail Infrastructure Managers (EIM) and UIC, representing the railways, present a joint roadmap for digital railways highlighting the opportunities and challenges of rail digitalisation. It is clearly highlighted that the objective for the rail sector is to offer highly efficient and attractive transport options to their customers and to make the most of the opportunities offered by digital transformation.

With this roadmap, the sector proposes some objectives to make railways digital by:

- Offering connected railways by providing reliable connectivity for safe, efficient and attractive railways;
- Enhancing customer experience by offering better and added value for customers;
- Increasing capacity by enhancing reliability, efficiency and performance of railways;
- Boosting rail competitiveness by making the most of transport data.

First of all, provision of connectivity across the railway network could answer, from the operational point of view, to the strong need for highly available, reliable and stable network connectivity, while meeting the technical, operational and functional requirements of the railway system.

The development of digital tools could also address individual requirements by creating door-to-door solutions and added value for the customer experience before, during and after travel. For instance, to respond to the increased demand from customers, it would allow having the ability to access the Internet and then to enhance customers' experience.

The emergence of Internet of Things will help to increase productivity and effectiveness of operational processes. The challenge for railway sector will be to implement new sensors on existing equipment, and to create, collect and make the best use of data in an efficient way while ensuring security and privacy of data.

Finally, the development of ICT has facilitated the collection and exploitation of transport data. The roadmap highlights that maximising the use of data will allow to boost rail competitiveness, leading to economic growth, innovation and significant benefits for the rail sector, creating and developing interoperable and interconnected services.

Nomad Digital, a leader in on-vehicle ICT solutions, as presented in Chap. 2, performed a rail industry survey on the major challenges that rail operators, maintainers and owners have to face to remain competitive and the role of ICT to face these challenges [45]. Currently, the challenges for railway operators are numerous, from the more efficient operation, the maintaining safety and service quality, the more complex modern rolling-stock to manage, the new forms of competition, the increasing cost of energy, the demanding standards for emissions and the increase of ever-demanding passenger needs in terms of punctuality, information, amenities and safety. Globally, it consists in achieving more with reduced budget.

The survey highlights that the main concerns for rail industry are the operational and maintenance costs, and the providing of high quality service and information to the passengers. Nomad Digital argues that "connectivity" and on board networking of the many on board systems is the key to solving many of the challenges. Transport operators are starting to recognize the value that connecting vehicles can bring. However, it is crucial to bring support in understanding the benefits of new technologies.

The challenge is to reach a "networked train" philosophy. It consists for the several different existing systems to share data, self-analyze and share with other systems, operations centers and maintenance depots to allow operators to make intelligent decisions and optimize their operational activities, from maintenance to dynamic passenger information and lower energy consumption.

The Fig. 3.5 summarizes all the applications that could be included in the "networked train". It consists in:

- Real-Time passenger infotainment systems: real-time status of passenger's journey, information on disruption and entertainment;
- Telemanagement: gathering maintenance methodologies and condition monitoring;
- Ecodriving and energy monitoring: optimization of driving techniques and reduction of energy consumption;
- Real-time CCTV: live management of security events and internal and external train environment;
- Automated passenger counting: dynamic indication of train load for passengers and operation centers;
- Passenger Wi-Fi: Internet on board train for passengers.

In [46], authors highlight that railway domain have to satisfy its customers to be competitive. To do so, innovative technologies have to be integrated into the existing systems to improve efficiency and performance. A panel of innovative railway industries was studied, highlighting the different way to innovate in the domain.

In [43], the authors confirm the need and the added value of the radio communications in the railway domain. They even talk about "disruptive technology". Indeed,

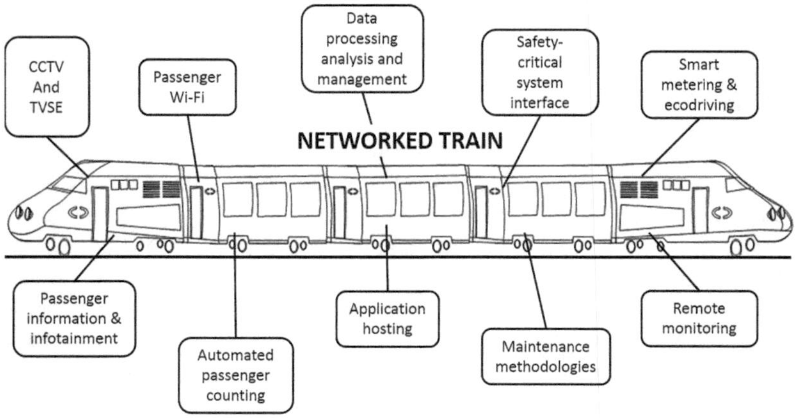

Fig. 3.5 Illustration of the "networked train", inspired from [45]

radio communication technologies represent a flourishing field and can improve efficiency, safety and profitability. Several technological issues are summarized, such as the radio convergence, the withdrawal of on board wiring, the High Speed scenarios or the channel modelling. All these aspects were detailed in this book.

References

1. Spectrum policy task force. Technical report, Federal Communications Commission, ET Docket No. 02-135, November 2002.
2. General survey of radio frequency bands (30 MHz to 3 GHz). Technical Report 20100323, Vienna, virginia, Shared Spectrum Company, September 2010.
3. Joseph Mitola III and Gerald Q Maguire Jr. Cognitive radio: making software radios more personal. *Personal Communications, IEEE*, 6(4):13–18, 1999.
4. Jacques Palicot. *Radio Engineering - From Software Radio to Cognitive Radio*. Wiley, Paris, October 2009.
5. S. Chaudhari. *Spectrum sensing for cognitive radios: algorithms, performance, and limitaions*. PhD thesis, Aalto University, 2012.
6. Erik Axell, Geert Leus, Erik G Larsson, and H Vincent Poor. Spectrum sensing for cognitive radio: State-of-the-art and recent advances. *Signal Processing Magazine, IEEE*, 29(3):101–116, 2012.
7. Linda E. Doyle. *Essentials of Cognitive Radio*. Cambridge University Press, 2011.
8. Piotr Palka. Future terrestrial broadcast systems: An overview of the use of TV white spaces. *Consumer Electronics Magazine, IEEE*, 2(3):17–24, 2013.
9. Amit Kumar Mishra and David Lloyd Johnson. *White Space and Communication - AdAdv, Developments and Engineering Challenges*. Springer, 2015.
10. http://grouper.ieee.org/groups/dyspan/.
11. http://www.cost-terra.org/.
12. http://www.cost.eu/COST_Actions/ict/Actions/IC0902.
13. Ashwin Amanna, Manik Gadhiok, Matthew J Price, Jeffrey H Reed, W Pam Siriwongpairat, and T Kee Himsoon. Railway cognitive radio. *Vehicular Technology Magazine, IEEE*, 5(3):82–89, 2010.

14. http://www.ict-phydyas.org/.
15. http://www.ict-qosmos.eu/.
16. http://www.ict-emphatic.eu/.
17. http://corridor.ifsttar.fr/.
18. Karim Hassan, Roland Gautier, Iyad Dayoub, Marion Berbineau, and Emanuel Radoi. Multiple-antenna-based blind spectrum sensing in the presence of impulsive noise. *Vehicular Technology, IEEE Transactions on*, 63(5):2248–2257, 2014.
19. Ali Kalakech, Marion Berbineau, Iyad Dayoub, and Eric Pierre Simon. Time-domain LMMSE channel estimator based on sliding window for OFDM systems in high-mobility situations. *Vehicular Technology, IEEE Transactions on*, 64(12):5728–5740, 2015.
20. Sofiane Kharbech, Iyad Dayoub, Marie Zwingelstein-Colin, Eric Pierre Simon, and Karim Hassan. Blind digital modulation identification for time-selective MIMO channels. *Wireless Communications Letters, IEEE*, 3(4):373–376, 2014.
21. Kamal Deep Singh, Priyanka Rawat, and Jean-Marie Bonnin. Cognitive radio for vehicular ad hoc networks (CR-VANETs): approaches and challenges. *EURASIP Journal on Wireless Communications and Networking*, 2014(1):1–22, 2014.
22. Jong-Hyouk Lee, Keshav Dev Singh, Jean-Marie Bonnin, and Sangheon Pack. Mobile data offloading: a host-based distributed mobility management approach. *Internet Computing, IEEE*, 18(1):20–29, 2014.
23. Huawei. The second phase of LTE-Advance - LTE-B: 30-fold capacity boosting to LTE. Technical report, 2013.
24. 5GPPP. 5G Vision - The 5G Infrastructure Public Private Partnership: the next generation of communication networks and services. Technical report, White Paper, March 2015.
25. Afif Osseiran. The 5G Mobile and Wireless Communications system. In *ETSI Future Mobile Summit*, November 2013.
26. Wang Zhigin. IMT vision towards 2020 and beyond. In *ITU-R Working Party 5D Workshop - Research Views on IMT Beyond 2020*, Ho Chi Minh City, Vietnam, February 2014.
27. Junyoung Nam, Jae-Young Ahn, Ansuman Adhikary, and Giuseppe Caire. Joint Spatial Division and Multiplexing: Realizing Massive MIMO Gains with Limited Channel State Information. In *46th Annual Conf. Information Sciences and Systems*, 2012.
28. Erik Larsson, Ove Edfors, Fredrik Tufvesson, and Thomas Marzetta. Massive MIMO for next generation wireless systems. *Communications Magazine, IEEE*, 52(2):186–195, 2014.
29. Fredrik Rusek, Daniel Persson, Buon Kiong Lau, Erik G Larsson, Thomas L Marzetta, Ove Edfors, and Fredrik Tufvesson. Scaling up mimo: Opportunities and challenges with very large arrays. *Signal Processing Magazine, IEEE*, 30(1):40–60, 2013.
30. Clayton Shepard, Hang Yu, Narendra Anand, Li Erran Erran, Thomas Marzetta, Richard Yang, and Lin Zhong. Argos: Practical Many-Antenna Base Stations. In *International Conference on Mobile Computing and Networking*, Istanbul, Turkey, August 2012.
31. Xiang Gao, Fredrik Tufvesson, Ove Edfors, and Fredrik Rusek. Measured propagation characteristics for very-large MIMO at 2.6 GHz. In *46th Annual Asilomar Conference on Signals, Systems and Computers*, Pacific Grove, Canada, November 2012.
32. Jakob Hoydis, Cornelis Hoek, Thorsten Wild, and Stephan ten Brink. Channel Measurements for Large Antenna Arrays. In *IEEE International Symposium on Wireless Communication Systems*, Paris, France, August 2012.
33. Madhavendra Richharia. *Mobile satellite communications: principles and trends*. John Wiley & Sons, 2014.
34. Réseaux mobiles professionnels, état des lieux et besoins futurs en fréquences. Technical report, ARCEP, October 2012.
35. http://www.agurre.fr/.
36. iDate Consulting. Evolution of GSM-R. Technical report, European Railway Agency, April 2015.
37. J. Cellmer and D. Mandoc. Future Railways Mobile Communication Systems. In *12th UIC ERTMS World conference*, Brussels, 29 February - 2 March 2016.

38. Hassan Ghannoum and David Sanz. Internet Onboard: technical analysis. In *5th International Workshop on Communication Technologies for Vehicles, Nets4cars/Nets4trains*, Lille, France, May 2013.
39. David Sanz. Satellite Technologies for Broadband Internet Access Onboard High Speed Trains. In *7th World Congress on Railway Research*, Montreal, Canada, June 2006.
40. David Sanz, P. Pasquet, P. Mercier, Bernadette Villeforceix, and David Duchange. TGV Communicant Research Program: from research to industrialization of onboard, broadband Internet services for high-speed trains. In *8th World Congress on Railway Research*, Seoul, Korea, May 2008.
41. Shit2Rail. Shift2rail founding members on the Shit2Rail Master Plan. Technical report, Shitf2Rail, 2014.
42. Shift2Rail Joint Undertaking, Multi-Annual Action Plan. Technical report, November 2015.
43. Juan Moreno, Jose Manuel Riera, Leandro De Haro, and Carlos Rodriguez. A survey on future railway radio communications services: challenges and opportunities. *Communications Magazine, IEEE*, 53(10):62–68, 2015.
44. CER, CIT, EIM, and UIC. A Roadmap for Digital Railways. Technical report, 2016.
45. Nomad Digital. Rail industry survey - Major challenges facing rail operators, maintainers and owners and the role of ICT. Technical report, 2015.
46. SmartRail. Using SmartRail technology to build the railway of the future. Technical report.

Conclusion

Wireless radio communications represent a real additive value for the railway domain. It implies challenges and opportunities for present and future railway applications. All railway services are impacted, from the critical applications regarding train control to the services to passengers. LTE and future 5G technologies represent a good opportunity to provide a framework for radio convergence that would enable to offer all the services over a converged media, keeping high Quality of Service and ensuring required security levels.

Many works still have to be done to change mentalities and way to think in a railway environment that evolves very slowly. The objectives of reduction of costs, which became a priority these last years, will push more and more to rely on innovative radio technologies. The program and objectives fixed by the Joint Undertaking Shift2Rail will support all these new technologies to reduce costs, enhance capacity and interoperability, improve reliability, reduce environmental footprint and support competitiveness of the sector.

É. Masson and M. Berbineau, *Broadband Wireless Communications for Railway Applications*, Studies in Systems, Decision and Control 82,
DOI 10.1007/978-3-319-47202-7

Appendix A
How Does Internet Work?

Internet operation leans on the TCP (Transmission Control Protocol) / IP (Internet Protocol) protocols. TCP/IP provides end-to-end connectivity specifying how data should be packetized, addressed, transmitted, routed and received at the destination. This functionality is organized into four layers similar but prior to the OSI model, illustrated in Fig. A.1:

- The link layer is the lowest layer of the TCP/IP protocol. It regroups the PHY and link layers of the OSI model. It is the group of methods and communications protocols that only operate on the link that a host is physically connected to. The link is the physical and logical network component used to interconnect hosts or nodes in the network. A link protocol is a suite of methods and standards that operate only between adjacent network nodes of a local area network segment or a wide area network connection;
- The Internet layer is a group of internetworking methods, protocols, and specifications in the Internet protocol suite that are used to transport datagrams (packets) from the originating host across network boundaries, if necessary, to the destination host specified by a network address (IP address), which is defined for this purpose by the Internet Protocol (IP). Two main Internet Protocols exist:
  - The IPv4 protocol, for which addresses are coded on 32 bits, which can be public, that is to say contactable from any computer connected to the Internet;
  - The IPv6 protocol is an evolution of the IPv4 one. Addresses are coded on 128 bits, which can increase the number of available public addresses. Evolutions concern the routing protocols and the automatic configuration of IP addresses.
- The transport layer establishes a basic data channel that an application uses in its task-specific data exchange. The protocols of the transport layer can solve the problem of reliability of the exchange and ensure that the data arrives in the correct order. In the following TCP / IP protocols, transport protocols also determine to which application each data packet has to be delivered. The transport layer ensures

É. Masson and M. Berbineau, *Broadband Wireless Communications for Railway Applications*, Studies in Systems, Decision and Control 82,
DOI 10.1007/978-3-319-47202-7

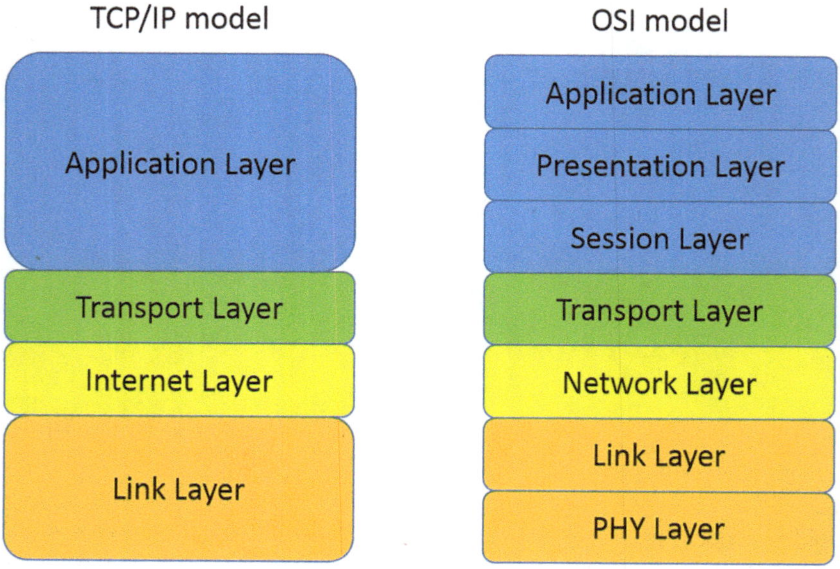

Fig. A.1 Description of the different layers of the TCP/IP model, compared to the OSI model

the following tasks: error control, segmentation, flow control, congestion control, and application addressing. The two main transport protocols are

- The TCP protocol is a "reliable" protocol, connection-oriented, that provides a reliable byte stream ensuring the data arrives unaltered and in order, with retransmission in case of loss, and elimination of duplicate data;
- The UDP protocol is a simple "non reliable" protocol, connectionless, which does not check that the packets arrived at their destination, and does not guarantee their arrival in the order. UDP is generally used for multimedia streaming applications or for applications based on simple mechanisms of question/answer.

The choice between UDP and TCP depends on the targeted application. If performance are more important than integrity, for example for a flow of IP telephony, UDP is favored. Conversely, when integrity is a primary request, TCP is chosen.

- The application layer includes the protocols used by most applications for providing user services or exchanging application data over the network connections established by the lower level protocols, but this may include some basic network support services, such as many routing protocols, and host configuration protocols. Examples of application layer protocols include the Hypertext Transfer Protocol (HTTP), the File Transfer Protocol (FTP), the Simple Mail Transfer Protocol (SMTP), and the Dynamic Host Configuration Protocol (DHCP). Data coded according to application layer protocols are encapsulated into transport layer protocol units (such as TCP or UDP messages), which in turn use lower layer protocols to effect actual data transfer.

The Domain Name System (DNS) is a hierarchical distributed naming system for computers, services, or any resource connected to the Internet or a private network. It associates various information with domain names assigned to each of the participating entities. Most prominently, it translates domain names, which can be easily memorized by humans, to the numerical IP addresses needed for the purpose of computer services and devices worldwide. The Domain Name System is an essential component of the functionality of most Internet services.

Appendix B
Mobile Satellite Systems

B.1 Architecture of a Mobile Satellite Systems

A satellite system is composed of the space segment and the ground segment:

- The ground segment consists of fixed earth stations (gateways), a network control center and operation control centers. These two latters manage the entire network, the satellite operation and the orbit control. Gateways behave like network interfaces between several external networks and the satellite network. They also perform protocols, addressing and format conversions;
- The space segment consists of the set of satellites of the system, which provide connection between the users and the gateways. Direct connections between users via the space segment is also achievable if the latest generation of satellites and users terminals are being used. The space segment of an operator may have one or more constellation of satellites each with an associated set of orbital and individual satellite parameters to facilitate the use of certain services.

Satellites are often used in a "bent pipe" architecture. In this case, they behave like repeaters between two communication points of the ground. Other satellites authorize On-Board Processing (OBP), which can include demodulation/remodulation operations, decoding/recoding operations, etc. in order to optimize the channel efficiency. In the case of OBP, the system can have Inter Satellite Links (ISLs), which consist of connections between two satellites in LOS.

É. Masson and M. Berbineau, *Broadband Wireless Communications for Railway Applications*, Studies in Systems, Decision and Control 82,
DOI 10.1007/978-3-319-47202-7

B.2 Characteristics of a Mobile Satellite System

B.2.1 Frequency Bands

Satellites can operate at different frequency bands:

- L band: from 1 to 2 GHz;
- S band: from 2 to 4 GHz;
- C band: from 4 to 8 GHz;
- Ku band: from 12 to 18 GHz;
- Ka band: from 27 to 40 GHz.

B.2.2 Types of Satellites

Two types of satellites exist:

- GEO (Geostationary Earth Orbit) satellites: they are located at a height of 35700 Km above equator, which leads to high propagation delays (around 400 ms) and high attenuations. They are typically used at high frequencies (S, L, Ku and Ka bands), attenuation is especially important. GEO satellites are initially adapted to fixed communications with large antenna on the ground. Nevertheless, several mobile systems exist;
- Non-GEO satellites: two categories exist: the LEO (Low Earth Orbit) satellites, located at a height between 500 and 2000 km, and MEO (Medium Earth Orbit), located between 8000 and 12000 km. The advantage of this type of satellite is the low altitude, which allows lower data transfer delays. The drawbacks are the low coverage, which requires use of several satellites and a management of handover between the satellites.

B.2.3 Existing Standards

Different satellite standards have been developed:

- Global System for Mobile communication (GSM) via satellite: GSM is one of the most used cellular technologies. This terrestrial technology can be adapted to satellite case. The most known is the GEO Mobile Radio (GMR) based on a GEO satellite. There are two specifications: GMR-1 and GMR-2. GMR has the same characteristics than GSM in terms of radio coverage, reuse of frequencies, protocol layers, etc. Mobile terminals can be "dual-mode" to capture GSM or GMR;
- Satellite Universal Mobile Telecommunications System (S-UMTS): S-UMTS leans on UMTS 3G technology. It can be complementary to UMTS but also cover

areas technically difficult for terrestrial systems. S-UMTS frequencies are around 2 GHz, near the frequencies of UMTS. Throughput can reach 144 kbps;
- Digital Video Broadcasting Satellite (DVB-S): DVB-S is a technology used around the word for TV broadcasting by satellite. The technology allows carrying MPEG-2 video streams and encapsulating IP data flows into MPEG-2 frames. Thus, for downlink, a single DVB-S transmission carries data for all users (data for each users are multiplexed in MPEG-2 frames). All users receive the same overall signal and extract their own data (data flow of each user is seen as a different channel for the DVB-S decoder, in the same way that numerical TV decoder selects and decodes a TV channel in particular). DVB-S being designed for TV broadcasting, it was adapted to small size antennas;
- Digital Video Broadcasting Satellite Version 2 (DVB-S2): DVB-S2 is the 2nd generation of standard for radio diffusion transmissions by satellite. It can be used also for point-to-point transmissions (like Internet) by an dynamical adaptation of coding and modulation phases;
- Digital Video Broadcasting Return channel over system (DVB-RCS): DVB-RCS utilizes Multi-Frequency Time Division Multiple Access (MF-TDMA) as a multiple access technique. Data are transmitted by time intervals on several different frequencies. The system was designed to manage capacity on demand: several time intervals and frequencies are then used for communications requiring large bandwidth, and conversely when less throughput is required. DVB-RCS was designed for fixed applications. Studies are realized to adapt to mobile cases, using especially Ku and Ka bands. New standard of DVB-RCS is available for mobile applications: DVB-RCS+M;
- Satellite Digital Multimedia Broadcasting (S-DMB): S-DMB predicts a diffusion component by satellite for 3G mobile telephony. It allows distributing multicast/broadcast multimedia service, which can be provided also by GSM and 3G. The architecture consists of a GEO satellite and some terrestrial repeaters, co-localized on 3G base stations. Its objective is to cope the dense urban areas suffering of LOS signal blocking. Frequency bands are located in VHF and L bands;
- DVB Satellite to Handheld (DVB-SH): DVB-SH is a standard for mobile diffusion based on OFDM, in order to provide audio and video broadcasting. DVB-SH allows a large coverage by combining a satellite component and terrestrial repeaters in NLOS areas. DVB-SH completes terrestrial DVB-H. Terminals can be dual-mode with a reception in S band (2.2 GHz) for DVB-SH and a reception in UHF band for DVB-H.

B.3 Main Existing Mobile Satellite Systems

The main existing mobile satellite systems are presented in this part:

- LEO satellites:

- Iridium: it is the ancestor of all mobile satellite systems. It covers the entire Earth surface, including oceans and polar areas. The system leans on 66 LEO satellites and ISLs among satellites, which allow voice communications and low data rate transmissions. 5 gateways are located on the Earth surface;
- Globalstar is the second precursor, with Iridium, of mobile satellite systems. The system consists of 48 LEO satellites. 25 gateways are in service, each one covering around 2000 km. Globalstar provide real-time voice, data and fax. Maximum throughput is 9.6 kbps.

• GEO satellites:

  - Inmarsat leans on 12 GEO satellites providing mobile telephony, fax and data transmission on the entire Earth surface, excepting poles area. The most innovative system is the BGAN, which provides services such as telephony or Internet. It is used in plains, such as Airbus and Boeing;
  - Thuraya leans on two GEO satellites, leaning on GMR interface and covering 110 countries. Mobile terminals can be "dual-mode" to capture terrestrial GSM and Thuraya system;
  - Hispasat offers telecommunication services for military or civil purposes, broadcasting, broadband multimedia communication and many more. Hispasat leans on a GEO satellites constellation, which is able to support VoIP, P2P file exchange, video conferencing and real time applications. Hispasat uses the DVB-RCS standard [1];
  - SES Astra is one of Europe's first private satellite operators and owns the Astra series of GEO communication satellites that provide various services from TV to Radio and Internet to millions of households. In 2007, Astra2Connect service is launched. It provides broadband Internet services. Astra2Connect basically targets households in remote and rural areas that cannot get broadband access from landlines but can also provide temporary backhaul link to the Internet if needed [1].

B.4 Criteria to Compare Satellite Systems

The different criteria to characterize mobile satellite systems are presented in this part [2]:

- The type of satellite (orbit): GEO, LEO or LEO;
- The frequency band: S, L, C, Ku or Ka;
- The used modulation at PHY layer: QPSK, 16-QAM, GMSK (specified modulation for the GSM standard), etc.
- The Multiple Access Technique:
  - TDMA: Time Division Multiple Access;
  - FDMA: Frequency Division Multiple Access;

  - CDMA: Code Division Multiple Access;
  - SCPC: Single Channel Per Carrier;

- Satellite operation:

  - "Bent-pipe" architecture: it is a way for satellite to relay information: data are transmitted to satellites, which return on ground. The only treatment done at satellite level is a retransmission of signals;
  - OBP (On Board Processing): in this case, signal processing are directly performed at satellite level.

- Use of ISLs;
- Used standard: GSM-satellite, DVB-S, DVB-RCS, etc.
- Supported applications: real-time applications, VoIP, Internet access, peer-to-peer, etc.

References

1. Panagiotis Georgopoulos, Jose Andre Moura, Rafidah Md. Noor, Ben McCarthy, and Christopher Edwards. Theoretical and Practical Survey of Backhaul Connectivity Options. Technical report, Lancaster University, Computing Department, June 2010.
2. Paolo Chini, Giovanni Giambene, and Sastri Kota. A survey on mobile satellite systems. *International Journal of Satellite Communications and Networking*, 28(1):29–57, 2010.

Appendix C
Existing and Future European Mobile Technologies

See Table C.1.

Table C.1 Existing and future European mobile technologies

Generation	Acronym	Download throughputs (theoretical/practical/usual)
1G	Radiocom 2000	Analogic
2G	GSM	9.05 kbps
2.5G	GPRS	171.2/50/17.9 kbps
2.75G	EDGE	384/64 kbps/–
3G	UMTS	144 kbps rural, 384 kbps urban, 1.9 Mbps fix/–
3.5G or 3G+	HSPA	14.4/7.2/3.6 Mbps
3.75G or 3G++ or H+	HSPA+	21/10/10 Mbps
3.75G or H+ Dual Carrier	DC-HSPA+	42/20/10 Mbps
4G (3.9G)	LTE	300/60/30 Mbps
4G/4G+	LTE-Advanced	1 Gbps fix, >100 Mbps in mobility/–/–
4.5G	LTE-A	10 Gbps fix/–/–
5G	LTE-B	50 Gbps fix/–/–

É. Masson and M. Berbineau, *Broadband Wireless Communications for Railway Applications*, Studies in Systems, Decision and Control 82,
DOI 10.1007/978-3-319-47202-7

FSC
www.fsc.org
MIX
Papier | Fördert
gute Waldnutzung
FSC® C083411

Kolibri 360 GmbH
Ferdinand-Jühlke-Straße 7
99095 Erfurt, Deutschland
produktsicherheit@kolibri360.de